5 $\frac{40}{mo}$ + 60ph EN

D0501452

DATE DUE

GAYLORD			PRINTED IN U S A

THE
DRAMATIST'S EXPERIENCE

By the same Author

★

SHAKESPEARE'S TRAGEDIES

JOHN FORD AND THE DRAMA OF HIS TIME

THE JOHN FLETCHER PLAYS

THE DRAMATIST'S EXPERIENCE

*With Other Essays in
Literary Theory*

CLIFFORD LEECH

PN
1623
L4
1970 b

1970

CHATTO & WINDUS

LONDON

EERL Withdrawn
Surplus/Duplicate
PROPERTY OF
UNIV. OF ALASKA LIBRARY

Published in Great Britain by
Chatto and Windus Ltd
42 William IV Street
London WC2

First published in the United States of America, 1970
by Barnes & Noble Inc.

SBN 389 03992 6

© Clifford Leech 1970

Photoset and printed in Malta by
St Paul's Press Ltd

CONTENTS

PREFACE

THE writings in this book depend on the notion that criticism is, in its proper condition, agnostic, existentialist, republican. The critic can claim no definitive status for his judgment: he may be helped to form it by comparison between writing and writing, between characteristic modes operating at different times, between Kind and Kind, between writings belonging to different Kinds, but the judgment must be arrived at through an act of personal, non-logical commitment; such judgments, though in large measure they can be shared, will be primarily valid for a single person at one moment of time, a moment when mental engagement with the writing in question is at its most intense; and, without at all equating human powers of judging, the critic of the sort here implied will believe that every reader must be encouraged to judge as best he can.

It will be apparent that a datum throughout is that the critic's ultimate concern is judgment, evaluation, not classification into types.

The book is about theory, not criticism itself. Such judgments as are made are incidental and doubtless often cursory. The dominant view appears in outline in the first section, '"The Servants will do that for us"', where a stand is taken against the idea that the poet or maker as such is concerned with communication. Most of the remaining essays draw attention to special phenomena either recurrent in literature or associated with a particular place and time, but all of them may perhaps be useful in exploring certain basic features of writing. The last two essays touch, in general terms, on the special conditions of dramatic composition. Throughout there is emphasis on drama and the theatre's demands: it may therefore be appropriate that one of these final essays has been given prominence on the title page.

I owe great debts to the critics I have read and listened to (some of whom have been among my colleagues at Toronto and Durham), whether or not they are referred to in these pages (some names occur often), whether or not their views are at all close to those here put forward.

Certain sections have been printed before, and appear here with additions and in some cases with substantial revision of wording. 'A School of Criticism' was a lecture given at Durham and was published by the University; 'Comedy in the Grand Style' was the W. D. Thomas Memorial Lecture for 1965 and was published by the University College of Swansea; 'When Writing Becomes Absurd' was one of my George Fullmer Reynolds Lectures in the University of Colorado in 1963 and appeared in a pamphlet and also in *The Colorado Quarterly*; 'The Shaping of Time' was my contribution to *Imagined Worlds*, the John Butt memorial volume published by Methuen in 1968; 'Art and the Concept of Will', 'Elizabethan and Jacobean' and 'Congreve and the Century's End' appeared respectively in *The Durham University Journal, Queen's Quarterly* and *Philological Quarterly*. For permission to reprint these essays I am most grateful to the following: to the Council of the University of Durham for 'A School of Criticism'; to Professor Cecil Price and the Principal of the University College of Swansea for 'Comedy in the Grand Style'; to Professor Henry Pettit, Vice-President Eugene Wilson (chairman of the Reynolds Lectures Committee) and Professor Paul Carter (editor of *The Colorado Quarterly*) for 'When Writing Becomes Absurd'; to Methuen and Co. Ltd., and Professor Maynard Mack for 'The Shaping of Time'; to Professor Mervyn James (editor of *The Durham University Journal*) for 'Art and the Concept of Will'; to Professor H. Pearson Gundy (editor of *Queen's Quarterly*) for 'Elizabethan and Jacobean'; to Professor C. A. Zimansky (editor of *Philological Quarterly*) for 'Congreve and the Century's End'.

I am further grateful to Secker & Warburg Ltd., London, and New Directions, New York, for permission to quote from Italo Svevo's *Senilità* , translated by Beryl de Zoete, on pp. 144–5.

My wife has been characteristically gentle in her encounter with it all, characteristically discriminating in her comments.

Toronto, April 1969   CLIFFORD LEECH

Chapter 1

'THE SERVANTS WILL DO THAT FOR US'

T HE argument that follows involves conjecture on what
happens in a poet's mind and on the character of poetry
itself: to deal with such things is a rash undertaking which
the teacher or literary historian or critic might be expected
to leave more safely to the poet. Nevertheless, we cannot
be teachers of literature without making the attempt, in our
own fashion, to contemplate the nature of the poetic act.
Many of us to-day, moreover, feel an increasing need to
grapple with two related problems: the relation of the
poet to the society in which he lives and which is necessarily in his mind as he writes; and the relation of the poem
to mere vehicles of communication.

My starting-point is to be found in *Axël*, that splendid,
perhaps unactable, drama written in 1885 by Villiers de
l'Isle Adam. In the fourth and last part of this drama Axël
and Sara have defeated their enemies—for her the church,
for him men with a drive for power and with a pretension to
false wisdom—and there is now nothing to stand in the way
of their desire to explore life freely and passionately. They
are young, and in love; they have wealth and rank; the world
is open to them. They tell each other, in speeches that extend
over several pages, of the long travels they can undertake,
seeing all things in nature and art that the world possesses;
and they will respond, they think, more deeply and sharply
to these things because they are in love and will be seeing
them together. And when the talking is over, there is a
moment of revulsion. In imagination they have done all the
things they have spoken of, they have seen the whole world
in their mind's eye—in a moment of exaltation, a moment

1

never to be recaptured because never again will they know the special feeling of liberation from society's claims that they know now, never again will the moment return when they first found and admitted their love. If they set about putting the long programme into effect, the actuality will be a poor thing in comparison with what is happening now in their minds. Their whole shared life will be a vain attempt to get back to the moment that is passing. So Axël makes what he thinks is the logical proposal: in their moment of triumph and foresight they have lived more fully than is ever possible for most people, more fully than they too will live if they proceed to the muted existence that actuality offers; so let them die together now. 'Vivre? Les serviteurs feront cela pour nous': what actuality can offer to those who have known the full life of the imagination would merely reduce them to the rank of the underprivileged. The play thus ends with their shared suicide.

I am not concerned here with the rightness or wrongness of their choice. What I am concerned with is Axël's notion in the sense of its being the extreme term in a long series. Over some centuries poets and other makers have been fascinated by the contemplation of modes of behaviour which definitively repudiate what is commonly understood by the term 'living'. A comic counterpart to Axël's repudiation is to be found in a novel by a contemporary of Villiers de l'Isle Adam, the *A Rebours* of J.-K. Huysmans, written in 1884. Here the central figure Des Esseintes thinks of breaking out from his solitude in Paris and paying a visit to England. He has been reading Dickens, and this, we are told, induces 'visions of English life over which he sat pondering for hours';[1] then Paris happens to be suffering from rain and fog, which helps to fix the idea of England more firmly in his mind. He packs hurriedly, gets into a cab, and dreams of the fog-bound Thames and the 'terrible world of men of business', the 'ruthless machine grinding to

---

[1] Quotations are from the translation by Robert Baldick: *Against Nature*, Harmondsworth, 1959.

powder millions of the poor and powerless, whom philan-
thropists urged, by way of consolation, to repeat verses of
the Bible and sing the Psalms of David'. He goes into a
foreign bookshop to buy Baedeker's or Murray's guide to
London: he luxuriates in the strange sight of English-bound
volumes, the stranger sight of 'foreign females'. As he
turns over the pages of the guide-book, he thinks of Pre-
Raphaelite pictures, of the personality of G. F. Watts—
'the personality, at once complex and essentially simple,
of an erudite and dreamy Englishman, unfortunately haunt-
ed by a predilection for hideous tones'. Then he surrounds
himself with 'swarms' of English and American visitors in
a drinking-place they frequent; he orders an amontillado,
and thinks of Edgar Allan Poe. Still he is too early for the
next train leaving for the Channel: he goes to the station
dining-room and watches the English visitors eat. He has
a dinner of oxtail soup, smoked haddock, roast beef and
boiled potatoes, Stilton cheese, and two pints of ale followed
by porter. But then he remembers how formerly he had
wished to visit Holland: he had gone there, and been disappointed. And has he not now for several hours been, in
effect, in London? 'I have been steeped in English life
ever since I left home; it would be a fool's trick to go and
lose these imperishable impressions by a clumsy change of
locality.' So, like Axël, he refuses the risk of matching the
'real' London with the experience his imagination has given
to him. He goes back to his solitude in Paris.

Huysmans gives us the whole episode with plenty of
macabre fun, but of course he means every word of it. Des
Esseintes is solitary and neurotic; he wants to know the
world and what it can offer, but to enter the throng, to rub
shoulders, to mingle—these things are not only painful,
they are of far less worth to him than the response to a
vision that the imagination has contrived out of far-off
glimpses and the pages of books. We find in his story a grimly
comic pathos, while we are likely to feel wonder at the repudiation made by the proud Axël and Sara: their act is
paradoxical, that of Des Esseintes is of a piece with the

man's uniform neuroticism. Yet all three characters make the same choice—with, it appears, the approval of the writers. This, of course, was the period when Pater found in art the means of living most intensely: the 'hard, gemlike flame' was to be got primarily through solitary contemplation, not from rubbing shoulders.[1]

But we cannot write off such pronouncements as peculiar to the late nineteenth century, as oddities likely to be found only in the attic of an ivory tower. It will be well to remind ourselves of some similar repudiations that we can find in literature of a much earlier date. We may think of the stoicism that the Elizabethans knew so well in Seneca. He in his plays could show men in high place destroying each other and themselves, and in his choruses he could insist on the need for setting oneself apart from the world of high office and internecine struggle: the thing that mattered was to be king of oneself, and that could be achieved only if one stood aside, watching other men's public and futile attempts at royalty—futile because the royalty they attempted was a mere public thing and therefore destructive of man's sole good aim. Or we may think of Lucretius finding some profit in remaining on the shore and watching the ships tumble in the waves. Or of the long pastoral tradition, in which poets and, later, playwrights imagined a world in the countryside where the ordinary pressures of life did not exist, where the sadness and the danger were manageable, where in the end the gods or their human representatives were manifestly on the right side. It is true that pastoral writing is of a mixed order, frequently commenting on the known world by means of juxtaposing it with an imaginary one, that satire and allegory are possible within the tradition, that even the actuality of country life and the actuality of human passion are not altogether banished from the poet's fields and woods. Later on I want to say something about such minglings of fancy and fact, of withdrawal with simultaneous looking back, but the basic notion of pastoral is

[1] As made fully explicit in the 'Conclusion' to *The Renaissance: Studies in Art and Poetry* (1873).

evidently enough a repudiation. So, too, we can recall the extravagances of the Platonic Love cult in Charles I's time. This derivative from Renaissance Neo-Platonism, and perhaps also from the Renaissance friendship cult, had as its patroness Queen Henrietta Maria, a wife and mother. It was 'as if' thinking carried to enormous lengths. It was mocked at in its own time—by Davenant, for example, in his play *The Platonic Lovers* of 1635—but the mockery was possible only because of the strength of the cult. It did not present an actual or a possible way of living: rather, it presented an impossible but, it seemed, better way—in which the only love-contact would be a communion of minds. And if the absurdity of the idea had become evident in the postures of a mannered court, we should remind ourselves that its finest literary expression came much earlier, in Shakespeare's poem *The Phoenix and Turtle*, published in 1601.

The female phoenix and the male turtle-dove are consumed in a mutual flame; their duality is at the same time a unity, a 'compounded' and 'concordant one'. But it is in the exchange of eyes that the intercommunion is achieved, and the 'threnos' to the poem proclaims that they died

> Leaving no posterity—
> 'Twas not their infirmity,
> It was married chastity.
> (ll. 59–61)[1]

The poem is not simply a celebration, for the birds are dead, and with that death only approximations to 'truth' and 'beauty' can now exist in the world. The last line's phrase 'these dead birds' suggests futility as well as loss. Yet along with the futility there goes a sense of a triumph against life, a rejection of common ways, an embalming of a virgin love,

---

[1] Throughout this book, quotations from Shakespeare are from *William Shakespeare: The Complete Works*, ed. Peter Alexander, London, 1951; act-, scene- and line-references are to the Globe edition (1865).

which in idea is not immeasurably remote from Axël's more loquacious testament.

And if the poets have often expressed an admiration, a half-longing, for a world apart, the notion is prominent too in the best and most gracious piece of Renaissance criticism in English. Sidney's *Apology for Poetry* contrasted, as everyone knows, the brazen world of nature with the golden world that poetry can offer, and it is worth noting some of the characteristics that Sidney looked for in this golden world. Its persons would be unmixed, not marked by the flawed goodness or partially redeemable wickedness of men in history; its events would be strictly governed by laws of causation, in no way subject to the Fortune that in the actual sublunary world could operate with apparent arbitrariness; and its rewards and punishments would correspond nicely to desert:

> If the *Poet* do his part aright, he wil shew you in *Tantalus*, *Atreus*, and such like, nothing that is not to be shunned; in *Cyrus*, *Aeneas*, *Ulisses*, each thing to be followed: where the *Historian* bound to tell things as things were, cannot be liberall, without hee will be Poeticall, of a perfect patterne, but as in *Alexander* or *Scipio* himselfe, shew doings, some to be liked, some to be misliked; . . . And whereas a man may say, though in universall consideration of doctrine, the *Poet* prevaileth, yet that the Historie in his saying such a thing was done, doth warrant a man more in that he shall follow. The answere is manifest, that if he stand upon that was, as if he should argue, because it rained yesterday, therfore it should raine to day, then indeede hath it some advantage to a grosse conceit. But if hee knowe an example onely enformes a conjectured likelihood, and so goe by reason, the *Poet* doth so farre exceed him, as hee is to frame his example to that which is most reasonable, be it in warlike, politike, or private matters, where the *Historian* in his bare was hath many times that which we call fortune, to overrule the best wisedome. Manie times he must tell events, whereof he can yeeld no cause, and if he do, it must be poetically. . . . for indeed Poetrie ever sets vertue so out in her best cullours, making fortune her well-wayting handmayd, that one must needs be enamoured of her. Well may you see *Ulisses* in a storme and in other hard plights,

6

but they are but exercises of patience & magnanimitie, to make them shine the more in the neare following prosperitie. And of the contrary part, if evill men come to the stage, they ever goe out (as the Tragedie writer answered to one that misliked the shew of such persons) so manicled as they little animate folkes to follow them. But the Historie beeing captived to the trueth of a foolish world, is many times a terror from well-dooing, and an encouragement to unbrideled wickedness.[1]

Very possibly some aspects of the revision of the same author's *Arcadia* are to be explained by a desire to bring its world closer to a golden condition: in the first version Pyrocles attempts to seduce Philoclea, and Musidorus comes near to committing rape on Pamela; Pyrocles and Musidorus had thus been presented as mixed characters, with the blemishes of actuality upon them; and for Sidney literature had come to be seen as needing to present a nobler or, rather, a simpler world. Certainly in the *Apology* he thinks partly in moral terms: by contemplating the golden world, we may be led to try to conduct ourselves as we should if we inhabited that world. Nevertheless, the imagined world can never be truly ours; it is one that, because we may be poets or readers of poetry, we can enter in imagination. But it is, Sidney makes no doubt, a better world than men can directly know.[2]

Indeed it is possible to find in literature many a tribute, or a wry glance of admiration, as the poet considers the manifestation of a life altogether remote from any that is open to us. Aldous Huxley, in his introduction to *The Letters of D. H. Lawrence*, commented on Lawrence's ability to appreciate the things of nature, not for any good they can do for us, not as in any way emblematic of our mode

[1] *The Prose Works of Sir Philip Sidney*, ed. Albert Feuillerat, Cambridge, reprinted 1963, III, 16–18. In the above quotation punctuation has been to some slight extent normalised.

[2] Elizabeth Dipple, 'The "Fore Conceit" of Sidney's Eclogues', *Literary Monographs, Volume I*, ed. E. Rothstein and T. K. Dunseath (1967), pp. 1–47, has argued that the verse eclogues in the *Old Arcadia* have the function of setting a picture of the shepherds' innocence against 'the moral abnormality of those who should be their ethical superiors'.

of living, but because they exist, because their world is not our world. Huxley proceeded to characterise Lawrence's wisdom as enshrining a 'doctrine of cosmic pointlessness', a feeling of delight in an otherness which has no relation to any problems we face.[1] We may remember how in Chapter XI of *Women in Love* Rupert Birkin exclaims at the grandeur the earth would possess if man did not exist to tidy and pollute it. Huxley himself has expressed a notion similar to Lawrence's in his own book *The Devils of Loudun*, where he refers to Christ's exhortation that we should consider the lilies, consider them 'not as emblems of something all too human, but as blessedly other, as autonomous creatures living according to the law of their own being'. He goes on to quote Whitman's poem 'Animals', where animals are admired because they have no share in the special properties of the human condition:

They do not sweat and whine about their condition,
They do not lie awake in the dark and weep for their sins,
They do not make me sick discussing their duty to God,
Not one is dissatisfied, not one is demented with the mania for
    owning things,
Not one kneels to another, nor to his kind that lived thousands
    of years ago,
Not one is respectable or industrious over the whole earth.[2]

Any way of living which is not ours may from time to time receive its tribute from the poets. The cloister, the army, the life at sea, the brothel—each of these in some measure removes the need to answer questions, to make decisions, to yield to the promptings of a noble or ignoble ambition.[3] 'Get thee to a nunnery. Why wouldst thou be a breeder of

---

[1] London, reprinted 1956, p. xviii.

[2] London, 1952, p. 94.

[3] The idea of such places as offering a refuge from the world seems particularly frequent in contemporary writing; as for example, ' . . . she was like a woman who, to feel secure at last, enters a brothel' (Georges Simenon, *Act of Passion* (*Lettre à mon Juge*), tr. Louise Varèse, Harmondsworth, 1965, p. 173).

sinners?' I do not think that Hamlet meant 'brothel' when
he said 'nunnery',[1] but he was expressing—and Shakespeare
was expressing through him—that desire to have done with
decision and involvement that led T. E. Lawrence to the
Royal Air Force. There is a passage in Graham Greene's
novel *Rumour at Nightfall* where one of the characters is
approached by a pimp, and suddenly sees the man as a kind
of ascetic: 'this that the man seemed to stand for was more
than lust; it was the ultimate desirability and ultimate beauty
of sterility. It was something worthy to be a creed and worthy
to be fought.'[2] The 'ultimate beauty of sterility' is something
Shakespeare could understand, in *The Phoenix and Turtle*
as well as in *Hamlet*, though he sensed the futility along with
the beauty.

More ordinarily, however, the impulse to withdraw into
another world shows itself in literature in the writings that
concern themselves with distant places and times. We can
make fun of the lower reaches of the historical novel, as of
the popular love-romance or tale of adventure, but these can
be seen as analogues to the more considered contemplation
of an otherness. One reason why modern dramatists re-tell
the stories of Greek tragedy, why Ingmar Bergman has on
occasion set his scene in medieval times and in one instance
(in *The Silence*) in an unknown country for which he himself
fabricated the language, why Akira Kurosawa has brought
the samurai to the cinema, is that through these devices the
nature of the human condition can be seen in isolation from
the random shoulder-rubbing of contemporary living.

It can be argued that all men experience from time to
time the urge to get away, to have done with day-to-day
decisions. But there is, I think, a special reason why the
poet, in the broad sense of the term, feels this urge with
greater force. As a poet, he must in a measure withdraw. We
may say that poetry is a full-time job. In older and more

[1] As J. Q. Adams and J. Dover Wilson have suggested: see the New
Cambridge *Hamlet*, reprinted 1961, pp. 193, 301.

[2] New York, 1932, p. 162.

respectable language, the poet 'ought himself to be a true poem'. For us that cannot mean, what it did mean for Milton, that his life is more noble, more 'golden', than other men's; but it does still mean that it is a life in which constant regard is given to the contemplation of existence, of its context, and of the mind that contemplates. It is a commonplace to say that John Donne watched himself experiencing the things that he wrote about, thus giving a special irony to his mode of documentation. But Donne's predicament is only a more obvious exemplar of the condition within which any major poet lives. We can see his predecessor Sidney looking at himself with a wry smile in *Astrophel and Stella*. We can sense Shakespeare shrugging at his own superlatives as he praises the Friend in the sonnets and berates the Dark Lady. And because of this impulsion to watch himself as he writes the poems, and indeed as he goes through the experiences that will contribute to the poems, the poet seems to himself as both separate from other men and painfully involved with them. He is in a measure distinct, yet he rubs shoulders; and he has a need to feel totally distinct. He cannot but long for an impossible condition in which there would be no ordinary contact, no need for everyday decision, no compromise.

Even so, the futility that Shakespeare hinted at in the flat words with which he ended *The Phoenix and Turtle* — 'For these dead birds sigh a prayer' — is powerful with other poets too. What, the writer feels bound to ask, is the advantage in being distinct yet never wholly distinct? Would it not be a better sort of life if one could forget distinctions, forget the need to watch oneself, forget the habit of poetic activity, and live as most men do? Perhaps, of course, no man lives in the totally impercipient way that the poet in such a mood envisages, but certainly the officially approved way of the world is remote from the poet's, and at times he will wish that it could be his. So Tennyson in *The Palace of Art* fears for the restrictions that art imposes on him; so Yeats wishes that a life of simple action had been, if it could have been, his choice:

'THE SERVANTS WILL DO THAT FOR US'

I turn away and shut the door, and on the stair
Wonder how many times I could have proved my worth
In something that all others understand or share.

But Yeats knows that the ordinary things are not for him:

But O! ambitious heart, had such a proof drawn forth
A company of friends, a conscience set at ease,
It had but made us pine the more. The abstract joy,
The half-read wisdom of daemonic images,
Suffice the ageing man as once the growing boy.[1]

The echo of *Macbeth* is, I think, significant. In his isolation
Macbeth too thinks of what he has lost:

that which should accompany old age,
As honour, love, obedience, troops of friends,
I must not look to have;

(V. iii. 24–6)

but, as Robert Ornstein has pointed out,[2] this is a fragile
good to set against the general evil of the world. No poet
could feel that this was enough for him, and we cannot
imagine Shakespeare's Macbeth as content even if he could
have possessed it.

For isolation has its peculiar rewards. The poet does not
live in a golden world; he is denied ordinary membership
of the brazen one; but he can glimpse a golden world, and
see the actual world more sharply because he does not fully
belong in it. He knows about 'birth, copulation, and death',
as Eliot's Sweeney simply put it; yet he has a surer sense
than most people have of a world apart from the world we
know; he will not usually claim that he has 'the philosophic
mind' that Wordsworth believed he had at an age of little
more than thirty, but he does know that the full range of
human living is more nearly available to him than to those
who are preoccupied with solicitude about the morrow or

[1] 'Meditations in Time of Civil War', *Collected Poems*, London, 1950,
p. 232.

[2] *The Moral Vision of Jacobean Tragedy*, Madison, 1960, pp. 233–4.

11

about the need to preserve the existent structure of society.

And, to the poet, the poem he makes will have some of the properties of the golden world, however far he may be from that world in the conduct of his life. The poem will be coherent, patterned, definitive, in contrast to the flux of things out of which it has grown, in contrast to the poet's own manner of existence, which has been half-in, half-out-of, the world of other men. So in a sense the poem will belong to a world apart from the world the poet has to live in. He will be aware that the poem's world is not his, that he cannot live with it, that indeed, as Keats put it, he himself is 'the most unpoetical of any thing in existence'.[1] His position can be compared with that of the architect who builds a palace or a church and, when the building is complete, is no longer free even to move within it. This was poignantly recognised by Goethe in a passage in *Elective Affinities*:

> In this connection, the architect's fate is the strangest of all fates. How often does he give his intelligence and his entire devotion to the creation of buildings from which he is excluded! The halls of royal palaces owe their magnificence to him, but he does not enjoy them in their full splendor. In places of worship he draws a line between himself and the Holy of Holies: he is no longer allowed to ascend the steps which he has built for the heart-stirring celebration—like the goldsmith who only from a distance worships the monstrance whose enamel and precious stones he has set together. It is to the rich man that the master-builder gives, together with the key of the palace, all the luxury and opulence he will never enjoy. Is it not obvious that in this way his art will gradually withdraw from the artist, when his work, like a child who is well provided for, need no longer fall back on its father?[2]

The separation of the poem from the world of non-poetic living is commonly emphasised by the use of immediately apprehensible devices within the writing. I. A. Richards

[1] Letter to Woodhouse, 27 October 1818.
[2] *Elective Affinities*, tr. E. Mayer and L. Bogan, Chicago, 1966, p. 165.

refers to the 'frame-effect' of metre as operating in this way:[1] we can think, too, of the adherence of a particular poem to a recognised Kind or to an established form like that of the sonnet; we can think of the Unities in drama as a means of separation from the flux, or of the free use of echo of, and allusion to, other writings which have already achieved that status of independent existence which the new poem, during the course of composition, is moving towards.

At this point it may be useful to make my position clear on two important matters. In relation to one of these there has perhaps already been enough implied for the purpose, but it will be as well to avoid a chance of ambiguity. I do not see the poet in the ideal condition as concerned with communication. Rather, he is making something which will allow him a sense of mastering an aspect or unit of experience, which will indeed provide him with momentary relief from the flux, and which will stand in his mind for his response to the particular unit of experience. The response in its final form will, of course, not be a mental condition immediately induced by an impact from the outside world: that immediate condition will be modified through the process of composition—through its relationship, that is, with the experience of craftsmanship—and it will be modified also through taking its place within a total body of experience already, in greater or lesser degrees of consciousness, in the poet's mind. But the sense of mastery can come only when the writing is truly separate from the poet, when he can think of it as a public, no longer a private, thing. The poem is thus available to others: and perhaps these others should include people without a close connection with the poet, for his friends may be able to fill in gaps from what they know of the writer and from the experiences they have shared with him, thus completing a process of composition which the writer has left unfinished. But availability is what matters: communication to some extent takes place, but it is not, I think, the poet's aim in so far as he is a

[1] *Principles of Literary Criticism*, London, reprinted 1934, p. 145.

13

poet; and it is not essential, for the sense of availability could be achieved for the poet even if his poems were never read.[1]

Yet in the case of major writings a reader is normally found. The work is encountered, and reacted to. It is formed out of the flux, becomes a coherent and detached thing, and then is absorbed into each reader's consciousness, often indeed substantially affecting, even moulding, the shape of that consciousness. The artifact, the self-subsistent thing, becomes part of a chain or a 'flowing', as Malcolm Lowry through his *persona* in the posthumous novel *Dark as the Grave wherein my Friend is Laid* has put it:

> he seemed to see how life flowed into art: how art gives life a form and meaning and flows on into life, yet life has not stood still; that was what was always forgotten: how life transformed by art sought further meaning through art transformed by life; and now it was as if this flowing, this river, changed, without appearing to change, became a flowing of consciousness, of mind. . .[2]

Later we shall have occasion to note that the responses of a multitude of readers can, it appears, affect the nature of the available artifact,[3] and this indeed may be implicit in what Lowry says. Nor can we overlook Eliot's contention in 'Tradition and the Individual Talent' that the character of each artifact changes as the tradition to which it belongs develops further: we see the poem differently because of all the poems that stand between it and us.[4] It is easy to deduce how nugatory is the idea of simple communication between mind and mind. A produces x, and x becomes a shaping element in B's consciousness, and x itself changes its condition because of its impingement on B and his many fellows, and because it is continuously taking a new place in an ever more complex tradition.

The second matter that needs a word here is a possible

[1] See below, 'A School of Criticism', pp. 30–2.
[2] New York and Toronto, 1968, p. 43.
[3] See below, 'On Seeing a Play', p. 197.
[4] *Selected Essays*, London, reprinted 1934, p. 16.

misunderstanding arising from the use I have made of
Sidney's notion of the golden world of poetry. He indeed in
this theory saw the world of poetry as an ideal creation
towards which men might aspire, and we are unlikely to
follow him in his demand for unmixed characters, strict
causation, and what has come to be termed 'poetic justice'.
And the reason why we cannot follow him is that in our
world of poetry we demand a representation of human
experience which we can accept as, in general terms, 'true'.
We do not object to simplification of human character (indeed we shall always get a measure of such simplification,
for only in a simplified form is a character apprehensible),[1]
and we neither insist on the idea of strict causation nor
strenuously object to it. Only by making things thus harder
in outline may a particular poet be able to gain control. We
shall object, however, if any property of the poem's world is
manifestly at odds with our own knowledge of related areas
of experience. This is to apply a truth-test, to talk of 'general
nature', and it remains the critic's ultimate function. It is
indeed a function that can be foolishly discharged. Not every
piece of writing is directed towards a total statement. There
are minor, exclusive Kinds—like comedy, pastoral, satire—
from which large areas of experience are eliminated. It is
perfectly legitimate to write a comic poem about a murder—
or about cancer, as J. B. S. Haldane did when he was dying
from the disease—or to write a novel about a small series
of love-affairs or a small process of growing-up without
reference to the Napoleonic wars in whose time—and Jane
Austen's—these small but valuable things happened. Where
exclusion is manifest, it is foolish to object to it, though we
should recognise the clear indication that the writing belongs
to a minor Kind. Within the limitations of that Kind, we shall
demand fidelity to the aspect of common experience dealt
with. And indeed we should remember that the poem cannot
become a public thing, cannot achieve separation from the
poet, unless it has this ring of truth: it is not 'available' to us

[1] See below, 'The Dramatist's Experience', p. 224.

15

unless we can in future accept it as contributing to our total view of things—which is not the same as saying either that it must fit in with our pre-existent views or that it will, without modification, be incorporated into our personal store of wisdom.

It must be remembered, too, that exclusion must be manifest. I expect that, after a lapse of some years since its publication, most of us would agree that Pasternak's *Dr. Zhivago*, impressive at it is, does not give a view of the Russian Revolution and its aftermath that we can accept as adequate to the total fact. That would not at all matter if the novel had been manifestly concerned only with the story of one man and his group of intimates: as far as that one man, one group, were concerned, the book is indeed admirable. But Pasternak's personal anguish had, it seems, made him unable to distinguish between the particular story and the historical context, and his book was presented as of the non-exclusive kind, as offering a total statement of a nation's predicament. Some readers have seen the central figure as 'placed', as the presentation of a man caught up in his time but seen, by the author, from outside it, but this interpretation seems at present hard to accept. It is for this reason that many of us are likely to feel inadequacy.

From all this it will be apparent that, though the poem achieves separateness, though it does not, in the ideal condition, have a design on us, it cannot be self-subsistent. It has come into being through the poet's need to objectify: it will not objectify unless not only he but we can accept it as an adequate representation (not, of course, a replica) of an experience that will be, because we are human, in some measure shared.

That is indeed the ideal condition. In practice there are many writings which we shall recognise as of major importance but which confessedly do have a design upon us. They will announce an intention, in Spenser's words, 'to fashion a gentleman or noble person in vertuous and gentle discipline' or, in Milton's, to 'justify the ways of God to men' or, as Jonson put it in his preface to *Volpone*, 'to inform men of the best reason of living'. It may well be that in many cases the process of composition cannot begin without an urge to communicate an opinion, to mould the prevailing attitude towards an aspect of living. Painters and musicians seem less embarrassed in this way, but writers, using the language of communication and persuasion, have almost inevitably a didactic turn of mind. If a poem achieves authority, it will, however, have broken free from the initial intention: we can, I think, feel this happening in *The Faerie Queene*, which has certainly become very much more than an educational treatise by the time it has reached its third book. And *Volpone* has perhaps little more than an incidental relationship to its preface. It may be noted in this instance that, after composition has been completed, the didactic urge may reassert itself for the writer, and this will be a sign that in some measure he regrets the separateness of the thing he has made: he feels an impulsion to bring it back into relation with his views on this or that, as so often Shaw did in the prefaces to his plays, or with a current or traditional notion of the poet's didactic task. In so far, however, as within the writing itself there is evidence of a design upon us, the work will not have achieved the condition of separateness that belongs to art.

Parenthetically, it ought to be recognised that a poet is not necessarily a poet in every one of his writings. He may at times use words for purposes of pure communication. He has opinions which, like the rest of us, he would like other people to share. So he writes about them as persuasively as he can. There is nothing wrong with that, but for the nonce he is concerned with communication, not with 'making'. Literature, in the broadest sense of the term, will include such essays in communication, but here I reserve the words 'poetry' and 'poem' for the writing that comes into full existence when it gets free from having a design upon us and achieves design in the sense of completed form.

Vestiges of a didactic purpose represent only one of the ways in which a poem may not fully succeed in achieving separateness. The poet depends on what he has experienced—this term of course including everything, from whatever source, that sets his mind working. But the poem made out of this experience is not to have the status of a personal confession or protest or lament. The poet does not unpack his heart with words, or wear his heart upon his sleeve—not while he is being a poet. The shame or indignation or grief that could be directly expressed in non-poetic communication is here part of an elaborately formed compound: it is not subdued, but related to a wide context of experience, including the poet's experience as a crafts-man. Within such a compound, indeed, it may win an authority which it could never have in direct expression. Tennyson's grief for Hallam is more apprehensible to us because it is now part of such a compound, although *In Memoriam* in certain respects has not become free, being held back by the poet's sense of mission. Frequently, however, a poem will be marred by traces of unassimilated emotion. This was perhaps the besetting weakness of the Romantics. When Wordsworth in the 'Immortality' ode assures us that 'A timely utterance gave that thought relief, And I again am strong', we may feel that he is too easily assuming our interest in his well-being. In recent criticism it has become customary to use the term 'I-figure' to refer to the *persona* a poet may adopt in lyrical writing. The convention is a useful one, but it should not blind us to the fact that the poet's 'I' not infrequently means himself. And such 'I-poems' may indeed stand on this side of separateness. It is well enough if, as characteristically in Donne, the poet watches the 'I', which is simultaneously both himself and a being observed by himself, but that situation is not to be assumed as present in all poetic utterance. Despite its dialogue-form, we may wonder if the *Epistle to Arbuthnot* is free from the note of unassimilated protest, if Pope does not too readily assume that we have no difficulty in forgiving him for his sureness of his own virtue. And we may feel uneasy about the conclusion of Wyatt's 'They flee from me, that sometime did me seek': the remembering here is splendidly done, brought into relation with a large body of experience, sensuously and passionately and with deep irony recalled, but when we reach the last line, 'I would fain know what she hath deserved', we may find the expression of simple indignation given too prominent a place. Our attention is forcibly directed to the particular experience of Sir Thomas Wyatt, which we are too personally invited to share.

But of course it is not only the poet who may err in thus keeping the poem tied to his body. Many a reader will fail to distinguish between poetry and communication, will look to poetry for moral guidance, for easily profitable doctrine. Shakespeare has inevitably suffered in this way, because the didactic tradition has made it seem proper that the leading English poet should especially offer a plan for living. In recent years this has made many commentators try to see him as a good Tudor establishment-man, offering wise thoughts on the need for stability in a troubled age, or as a man who can still be enlisted in the ranks of Christian preachers. For such commentators 'ambivalence' has become an unpleasant word, despite the fairly obvious fact that a total view must recognise contraries as co-present in most of the important human activities; and most of us will hold that he who claims certainty on a major issue is a man who slenderly knows himself. If we can accept the notion that poetry objectifies a developing experience in the poet's mind, and that the major kind of poetry will have related that experience to the total experience that the poet has previously made his own, we shall not look to poetry for a message, an order of the day. A major writer's total view is too complex to be reduced to the status of a 'message'. The reader may try to appropriate the poem, making it conform to his own chosen way of thinking or voting or worshipping. To some extent, doubtless, we all do this, but at least we ought to recognise that, in so far as we do it, we are making the poem shrink, are indeed making it into something other than the thing it is. A poem is—again I speak of an ideal condition—a thing apart from practical issues. It must

19

give us a sense of truth, but it will not tell us what to do.

It will be apparent that I have been concerned with exploring the separateness of the poem from both the poet and the reader, while at the same time insisting on the relation of the poem to a general experience which allows us to judge it and indeed demands that we shall judge it. But a special set of circumstances will operate when the poem in question is a play. No play is brought into being by a poet working alone. The actors, the directors, all the technicians of the theatre will be working along with the poet—whether he is in the theatre with them or whether he is dead or otherwise remote—and the play as performed will be necessarily a work of collaboration. Moreover, among the collaborators will be the audience, who will exert an influence on the play throughout the performance. At first sight it may appear that we have here too intimate, too sustained a relationship, on the one hand, between the writing and the poet (either in his own person or in the director as his representative) and, on the other hand, between the writing and the spectator: it seems extravagant to talk of separateness in such a context. Yet the very fact of collaboration will induce some degree of separateness: the poet remains important, but he is not the author of the total performance. The play, in fact, will have got partially out of his hands as soon as rehearsals have begun; even the influence of the audience will increase this separateness, for some element of unforeseen improvisation is essential to the theatre and will depend on the circumstances of an individual performance. However much the poet may say in rehearsal 'What I *meant* was that *x* should convey this or *y* that', he will feel his writing getting away from him. So, too, will the director, however dictatorial he tries to be. An author has frequently given objective expression to this sense of the play no longer being wholly his, by referring to himself in the third person in prologue, epilogue or induction, or even (as with Shaw and others) within the body of the play. Thus in Brome's *A Mad Couple Well Matched*, probably acted between 1637 and 1639, the prologue-speaker says of the poet:

> He's hearkening here, and if I go about
> To make a speech, he vows he'll put me out.

This prologue was apparently written by Brome himself, but he thinks of himself in the third person, as a man interested in the performance but by no means in full control of it. In earlier drama we have the discussion of the performance which Henry Medwall gave to the actors he called A and B within the play of *Fulgens and Lucres*, and the comments of Will Summers the jester throughout Nashe's *Summer's Last Will and Testament*: in both instances the author, by commenting on the play from within, has given to it a high degree of impersonality. We may see an extreme example of this phenomenon in N. F. Simpson's *A Resounding Tinkle*, where the author is brought on the stage as part of the entertainment, being made to admit that many things are happening which he in no way envisaged. Perhaps the type-situation with a new play is that the poet may first visit the theatre with a strong sense of personal property, but that by the time a performance, or perhaps even a rehearsal, is over he is aware that, royalties always apart, the play has taken on a separate existence. When it is published and he writes a preface, he may try again to assert a proprietorial right.[1]

The situation is more complex with the new relationship that an actual performance establishes, the relationship between the play and its audience. The exact nature of this will depend on the particular play, the policy of the theatre, the general attitude to dramatic performance current at the time. The ancient association of drama with ritual implies that, always to some extent, sometimes predominantly, the audience will see the performance as something enacted on their behalf, expressing their aspirations or fears or attitudes. When this situation becomes total, the play has not the character of a poem in the sense in which I have used the term: it is a joint communication, to the gods or to other

---

[1] The author's measure of continuing responsibility for his play is discussed below in 'The Dramatist's Experience'. pp. 225–6.

groups of people, asserting the audience's point of view. But such a situation is rarely total in a civilised context. The poet's work will, to a larger or smaller degree, strain at the limits of the audience's expectation, will present a total view that goes beyond what could be expected of a social group. In responding to the poet's words, the audience will quite customarily feel that their spokesman is exceeding the text that they would have given him. Not only Shavians went to see the early performances of Shaw's plays, and the Shavians too would often be taken by surprise. Euripides aroused opposition. We shall not, surely, be deserting the realm of probability if we assume that *Hamlet* and *Lear* induced more than a sober nodding of the heads. In our own day, it is not only socialists that go to see Brecht, and go more than once. That does not mean that the theatre may not offer at times the predominantly ritual act, or may not be used for the purpose of plain communication: governments have sometimes in this century used both theatre and cinema for the purpose of inducing popular support for a programme or policy. But the poet generally makes a dubious propagandist, and the work of major standing will escape from any kind of group-envelope.

Thus, however intimate the relation between audience and play (and it will indeed be intimate), the play of the kind which I include within the term 'poem' will preserve a remoteness: it will be something watched, something other than an expression of anticipated sentiment. And in this measure there is a sharp distinction between the audience and those who, as actors and technicians, have put the play on the stage. The performers and their associates have submitted themselves to the poem, and that is true despite the fact that, as we have seen, the fully achieved play is in part their own creation: during the act of performance, it is not their function to judge (though they may well do, should do, that before and after the performance): they exist for the moment to bring the play into the fullest existence of which they and it are capable. But the audience, while modifying the thing seen and heard, must all the time be judging it as we judge a poem we read.[1] And, of course, judgment is simply the recognition, or non-recognition, or partial recognition, of delight and enrichment. It is a singularly humane kind of judging that we practise, or ought to practise, in the realm of art. But because the work is judged it is separate.

We have travelled a long way from Axël, but we may recall that the imaginative structure which he shared with Sara was something other than, more splendid than, an actuality that the two of them might know. His dream of imagined experience I have taken as representing a fully achieved poem. We may believe that the making of poems is of importance in the world, that only by objectifying and thus separating their experience do men win a fugitive control over their condition; and we may believe too that by contemplating such modes of separation we who are not poets can in some measure share that winning of control. Axël left 'living' to his servants, and there is plenty for the servants to do. As we have seen, they will use language—for communication of plain fact or request or command, for an appeal to pity, for persuasion too, for argument such as I have been trying to put before you. It should be evident that I do not underrate the importance of Axël's servants, even though I may think that some users of language to-day receive wages beyond their deserts. But a good servant knows the limits of his capacity and function. A user of language for communication will do his job not only more efficiently but with something nearer integrity if he is aware of the things a servant cannot do, if he recognises the special status of the poem—a thing that has got free from one man and must not be enslaved to another, an embodiment of complex experience which will not tell us what to do but will extend and enrich our sense of the human condition, a gesture of rejection which is at the same time a fugitive winning of mastery.

[1] This matter of the audience's continuously changing and growing response is discussed below, in 'On Seeing a Play', pp. 197–203.

Chapter 2

A SCHOOL OF CRITICISM

THE title implies what I shall attempt to suggest should be the dominant aim of all who take up the critical task and what appear to be the chief problems that we meet in trying to realise that aim. Perhaps what is said will bear some relation to the criticism of all literatures and of the other arts. But the problems I shall touch on are more noticeable with English writing, because our own literature is written, for the most part, in language that has some appearance of familiarity and thus more immediately challenges an aesthetic judgment. During this century scholars in the humanities have rightly become ever more conscious of the need for research: we accumulate our learned periodicals, our recurrent bibliographies; we begin, from a respectful distance, to emulate our science colleagues in urging the expenditure of money on visual and mechanical aids, notably microfilm-readers and tape-recorders and collating-machines; we demand for the award of higher degrees a minute chronicling of literary-historical fact and an exploration of hitherto overlooked by-ways; we are learning, very slowly, how to edit an English classic. Nothing that is said here should be taken as suggesting an insufficient regard for these activities. They are indeed essential if our criticism is to be more than whimsy, more than the interesting or casual outpouring of an uninformed mind. The critic is always dependent on the ever-unfolding story of scholarship. Time and again our views of the meaning and value of a piece of literature must be revised in the light of scholarly discovery, scholarly tabulation.

But central to the present argument is the belief that English scholarship is ultimately justified in its usefulness to the critic, together with the belief that the best English scholar will not only be aware of this but will cultivate within himself the habit of literary judgment. There have been notable English scholars to whose more austere labours we owe an immense debt but who in their occasional ventures into literary criticism have shown the *naïveté* and the prejudice of any tea-table autocrat. There are writers of Ph.D. theses who have read much and diligently and have tabulated their findings, but have not profited in any increase of ability to understand or evaluate a piece of creative writing. Yet if the formal study of English literature is to be justified, it is for the sake of a deeper understanding, a surer foundation for judgment. And scholarship itself will suffer if the scholar is not deeply aware of the kind of thing he is examining. In fact, when we are considering the relation to one another of a Good and a Bad Shakespeare Quarto, when we are reflecting on the implications of the word 'Gothick' for the eighteenth century, when we are comparing the earlier and later versions of *The Prelude*, when we are exploring the history of the provincial playhouse in the days when it most notably flourished, we must remember that what we are doing is a necessary preliminary to the clarification of an aesthetic judgment. Otherwise our work will be only a hand-maiden to History, possibly useful to others but of small profit to ourselves.

The danger of divorcing literary scholarship from literary criticism is balanced by the danger that we shall base our literary judgments on too superficial, too widely diffused a knowledge of books. Matthew Arnold was deeply aware of the importance of studying English literature within a European context. Only in that way, he saw, could we avoid the 'provincial' judgment and see how our literature has reflected the interplay of native and foreign influences. And since Arnold's time the field that we have to survey has become much greater: to both east and west of us literatures have grown and have provided illuminating points of reference for the judgment of English books. We cannot comment on any nation's drama in the last fifty years without relating it to the work of Chekhov; we seriously limit our view of the modern English novel if Melville is not somewhere in our mind's eye. There is a danger that, because of this extension of our field of reading, we shall become skimmers of literature's cream, equipping ourselves for moderately well-informed conversation lacking in scholarly roots, in the basis for deep understanding. To set up as a critic, we need a wide basis for comparative judgment, but this must not prevent us from devoting a more special attention to a section of literary activity where we shall ultimately feel almost at home. In developing our field of knowledge we need to select our special building-sites, while sharpening all the while our consciousness of the whole landscape within which our structures will stand.

We must remember too that literature cannot be fully studied in isolation from the other arts. If we know something of the painting and music and architecture of a society, we shall understand better its literary impulses and modes of expression. In our own day literature and the film have often come close together, and we shall be inadequate critics of Graham Greene, for example, if we neglect his collaboration with Carol Reed.

But these difficulties arise only from the extent of productivity in literature and the other arts. They would not exist if we had world enough and time, and we can cope with them reasonably enough if we assign judicious proportions to our specialisation and our general regard for the human scene. A problem harder for us to confront is that of deciding the limits of literary criticism as a separate study. The critic is ultimately concerned with the making of judgments: he must be prepared in the end to say whether or not the work under consideration is 'a good book'. And in making such judgments he inevitably approaches the territory of the philosopher. The writing or reading of a book or poem is an activity and an experience: the value we associate with it cannot be divorced from the value we associate with other activities and experiences. If we pretend that the aesthetic has no relation with the ethical, we reduce the artist to the level of an agreeable decorator of the living scene. Yet it is part of our common experience that the reading of certain books has exercised as potent an influence on our lives as anything that has happened to us. For many, a reading of *King Lear* is important in the same way as the getting to know another human being may be important: the one may matter more or less than the other, but they will matter in the same way. Moreover, not only are literary values inseparable from other values, but the writer of the more ambitious sort is evidently concerned with displaying in his work a vision of totality and an attitude towards it. He is saying in effect: 'Here is my universe, and here is my place in it.' We are bound to ask ourselves whether we are qualified to judge such a literary presentation of the cosmos without being trained philosophers, without being as familiar with Plato as we are with Shakespeare. And if literary criticism thus impinges on the territory of the philosopher, we can hardly afford, either, to be ignorant of psychology. If we are fully to understand a piece of writing, we need to know as much as possible about what happened in the mind from which it came, about what happens in our own minds as we read. We may be pleased or awed or revolted, and there may be dark explanations of these phenomena. Can we, in arriving at a judgment, afford to base it on feelings and rationalisations whose nature and origin we have not explored? There can indeed be little doubt that we shall be better critics if we know something of these other disciplines, though even then, of course, there is the danger that, unless the imperfection of our knowledge is kept strenuously before us, we shall be led into palpable error and absurd simplification.[1]

Nevertheless, the literary critic must not abdicate, handing over the task of interpretation to the psychologist and that of judgment to the philosopher. Like the creative writer himself, he occupies a special position. The author of a poem,

---

[1] The influence of I. A. Richards' *Principles of Literary Criticism* will have been strongly evident throughout this paragraph.

though his work will imply an attitude to experience, is by no means necessarily a constructor of systems, an inquirer into basic principles of conduct or thought, or an analyst of behaviour. Rather he is a man impelled to make something, which will represent for him some part of what he has deeply known. As we have seen in the opening section of this book, from this act of making he will gain a sense of control over this experience, because he will have contained within a measurable space the flux of his feeling, the often contradictory impulses that he is aware of in himself. This will apply both to the whole work and to a particular part of it. In the creation of a fictitious character, especially of one in which the writer reflects a good deal of his own personality, there is this sense of mastery and release. Our consciousness of ourselves and of the world about us is resistant to pattern: the sum total of a moment of awareness is oppressive in its complexity and in its mutability. We must simplify and concentrate our attention to have any grasp at all, and we know that the process of time will deeply modify the present perception. But a fictitious character has a beginning and an end, a scheme of conduct, a whole plan of life which we can hold in our mind's eye. Because we can see the whole within an instant, it is proof against the flux of things and it has not the stultifying complexity of the actual. Byron, melodramatically but perceptively, expresses this idea in the Third Canto of *Childe Harold's Pilgrimage*. In this poem he had created the figure of Harold, so manifestly a self-projection that the reading-public refused to see a distinction between the character and the poet. In the end Byron was prepared to admit at least a resemblance, but he was with good reason reluctant to drop the character from the scheme of the poem. In the following stanza he speaks of the way in which he can apprehend the imaginary figure, as he cannot apprehend himself:

> 'Tis to create, and in creating live
> A being more intense that we endow
> With form our fancy, gaining as we give

The life we image, even as I do now.
What am I? Nothing: but not so art thou,
Soul of my thought! with whom I traverse earth,
Invisible but gazing, as I glow
Mix'd with thy spirit, blended with thy birth,
And feeling still with thee in my crush'd feelings' dearth.

(III, 6)

This sense of release from the flux of time, of keener apprehension, will be more fully available to the writer through the making of a whole work. To put it harshly, perhaps, Byron was so obsessed with himself when he was writing *Childe Harold* that it was the character rather than the poem that gave him this satisfaction. He outgrew this stage, as a major writer must, for straightforward self-projection depends on an inflation of the ego. We do not identify King Lear with Shakespeare, or Shatov in *The Possessed* with Dostoyevsky, or Jean Macquart in *La Terre* with Zola: it is this play and these novels as wholes that we can recognise as the symbols of the writers' apprehension. The writer with a highly developed consciousness will be preoccupied with the world around him, as it presents itself to his personal vision, rather than with the attractions of his own personality. The work as a whole, then, will be an image of his relation to experience. It will be a human being's attempt to get into focus his impressions, his sufferings, his pleasures, his standards. It will be his report, the report of an unusually sensitive mind, on what his experience, during the time of composition, has meant to him. The critic's concern is to apprehend that report and to judge its value.

This image of a man's experience and resultant attitude will in fact challenge our evaluation in the same way as the man himself and his behaviour will. As literary critics, in fact, we judge writings as we judge men. This will be the clue to our special approach. When we judge our friends or the people we casually know—we will, and we must, perform such acts of judgment—we do not demand an unvarying intellectual consistency from them or great learning or conformity with a specific ideology. Rather, we

29

wish to find a vitality, a resilience, an awareness, a capacity for suffering and for delight, an ultimate reserve. These qualities that give distinction to human beings we shall seek also in their works: they are the qualities that will make writing live, as they make some men out of the common.

Thus the critic will illustrate in his special task the kind of judgment that every man makes from day to day. He will be set some way apart only by his greater knowledge of writings of diverse sorts, like a man who has travelled much and won acquaintance with a multiplicity of human beings and societies. Moreover, he will have schooled himself to the act of judgment, to the clear-sighted yet warm appraisal: he will have learned the special pleasure that comes from keeping the eyes wide open. No one would suggest that we should form no opinion about our friends before consulting our philosopher and psychologist colleagues.

Works of art, we have seen, are entities which have come into the world as a result of an artist's attempt to win a sense of mastery over a portion of experience. They also make use of a more or less intelligible language. This is partly because, within a given community at a given time, its members will inevitably shape things for themselves in an approximately similar fashion. Most of us are bound to think and feel in terms that have a general currency. Partly, also, the intelligibility is due to a psychological need of the artist. In order to master his experience, he must give it a shape that will not be wholly private to him: he must accomplish a divorce between it and himself, so that he can contemplate it from outside. In other words, the art-work will become a public thing, which will stand for the artist's experience, in the eyes of a community as well as in the artist's private vision. Of course, to this there are apparent exceptions. There is the artist like Blake, with an intensely peculiar vision, which is given shape in a pictorial and literary language difficult of access. And there are those artists who are impelled to extend the frontiers of their language, finding the current idiom inadequate to their purpose. In their case the art-work, if it captures a general interest, will ultimately cause an enrichment of the current idiom. Nevertheless, we can say in large terms that the work of art is not only an entity but one that is valid for others besides the artist. Yet it is not a mere channel of communication, a means of transmitting thought and feeling from, say, a writer to his reader.[1] No reader can duplicate a writer's experience. On the one hand we have the experience of making, of contriving a new entity in the universe, and of gaining a limited mastery over the flux of consciousness. On the other hand we have the experience of contemplating the new entity, of coming to terms with it as with a situation or a human being that has newly entered our lives. We may, as readers, feel a deep sympathy with the writer, we may—if we are perceptive enough—glimpse something of the formlessness out of which form has been made. But works of art are not letters or intimate confessions addressed to us. They do, in the result, contribute to the sum of human sympathy, for the reader will have a sense of indebtedness to the writer, and, the work of art being a public thing, it will link together all those on whose consciousness it has impinged. Nevertheless, in our theory of art we should think of three entities: the artist, his work, and the member of his community who comes into relation with that work. A too close sense of relationship between the writer and the reader may make full contemplation of a piece of literature more difficult, for the reader will wish the writing to fit satisfactorily into an already established personal relationship and this desire may lead to distorted interpretation and biased judgment. There are of course occasional verses addressed to specific readers, where these considerations apply only in a modified way. But, while no one would wish to discourage the writing of such personal messages—they can be one of the graces of life—they are not, in the full sense of the term, works of art: unless the addressing of an

[1] See above, ' "The Servants Will Do That For Us" ', pp. 13–14.

individual is a politeness or a subterfuge, the writer is primarily aiming not at the making of something separate from himself but at the establishment of a closer contact with another particular human being.

It follows, obviously enough, that the study of a writer's life is irrelevant to the judgment of his work. It may help us in the task of interpretation, for many details of expression are likely to depend on private matters. But it may also be a hindrance even in this, for a writer's biography may arouse false expectations of the nature of his work. While, therefore, not neglecting any available information concerning what happened to a writer apart from his writings, we must not have too much confidence in this aid to interpretation. Above all, we must not confuse our judgment of the work with our judgment of the man. Each is an entity, challenging our separate opinion.

If we are not to see works of literature as documents of a writer's life or as personal messages haphazardly disseminated, we must face the problem of determining what qualities give value to a work. Here we need to distinguish between qualities of temper and qualities of structure. I have already suggested that our reaction to a work of art will be partially determined in the same way as our reaction to a human being: in language only on the fringe of metaphor, it is possible to speak of vitality, resilience, awareness, capacity for suffering and delight, and the sense of an ultimate reserve, as characteristic of a painting or a piece of literature. These things, you may say, reflect the mind of the painter or writer, but they may inform his work without being notably apparent in his personality. Our judgment of these things, however, cannot be subject to rule. All one can say is that we are more likely to judge wisely if our acquaintance with literature and art is wide, and if we approach any new work with inquiring and not fully committed minds. One cannot demonstrate, in a laboratory fashion, the vitality of *The Way of the World* or *The Ambassadors*: one can feel astonishment at the reader who fails to see it. If we would train a critic, at least we can urge wide reading upon him, and attempt to dispel that defensive arrogance that is the mark of the novice.

We can more easily, however, demonstrate qualities of structure. It is these that I wish to refer to in the remainder of this essay. I shall, of course, make no attempt to be exhaustive, but have chosen three aspects of a work of literature that seem important in their bearing on evaluation. These are: the scale of the work, its unity, and the degree of complexity within that unity. The first of these may be of particular relevance to English studies to-day. Largely as a result of the work of I. A. Richards, we have in our English teaching developed a method of dealing with the short poem or extract: we can scrutinise it phrase by phrase, and consider the way in which the words impinge upon one another. The danger is that we shall develop a strong critical awareness of the work of small compass, and fail to recognise fully the importance of large-scale proportions. This danger is all the greater because in our time the long poem is a rarity. Ever since Eliot performed his feat of condensation in *The Waste Land*, one has the impression that for some poets it seems hardly respectable to fashion anything of length. The splendid eccentricity of Pound's *Cantos* has been too private an exercise to encourage emulation. And when we are dealing with the longer works in our teaching, we are too often betrayed into vague description and loose generalisation. We need, in fact, a technique which will make plain the virtues of large-scale writing. For there can be little doubt of its special virtues. Repeatedly in *The Poetics* Aristotle stresses the importance of magnitude. Its acquisition was, for him, a decisive stage in the development of tragedy. Certainly he recognised that a work of literature should not be so long that we cannot hold it in our minds as an entity. But, provided that that condition was met, size was for Aristotle a virtue.[1] He had good reason for this view. When we are confronted with a work of some size, our

---

[1] Chapters VI, VII.

engagement with it is necessarily long. It will be weeks before we finish *The Faerie Queene* or *War and Peace*. Living with them for a while, we shall have the experience of becoming domiciled in their territory, of being able to move about within them. And because the reading of such books will provide for us, over a substantial period of time, a different mode of existence from that which we have in our non-reading hours, we shall have the sense that a whole cycle of life is within them. Here, we shall feel, is an image of totality. The writer has set out to master his experience as a whole, or at least as much of that experience as he can apprehend during the period of composition. The shorter work will give us the momentary abstraction, the joy or anguish of a single day. But if a piece of writing reaches a certain length, and if we feel within it a continuing process of growth from its beginning to its necessary end, we shall also feel that it corresponds to, and yet in its patterning transcends, our private image of a human life. No one would suggest that length in itself bestows high quality, yet the writer on the grand scale is attempting a more significant task than the writer of the short story or the lyric. We may note, by the way, that in poetry the sense of magnitude, of imaging a totality, may be given in a smaller compass than in prose: *Beowulf* has little more than three thousand lines, but the authority of its utterance is such that it gives the impression of large proportions; our published text of *The Waste Land* has little more than four hundred lines, but it has the dimensions of a major work. The novel, with its more casual manner and with its generally greater suitability for rapid reading, generally needs far more space to reach this level of implication; yet in *Heart of Darkness* and in *Billy Budd* the magnitude is achieved with economy. What matters in this regard is that we feel we have been long engaged with what we read: our major experiences in literature will indeed be with works of a certain magnitude.

The idea of unity in a work of art is inherent in the idea of its separateness. But it is evident that the form and degree of unity will vary from instance to instance. I have mentioned already the importance to Byron of the central figure in *Childe Harold's Pilgrimage*: in the early cantos of the poem the continuing presence of this fictitious character is the element that links the separate portions of the writing and enables us to hold the work in our minds as an entity; when in Canto IV the Childe disappears, there are more tenuous threads that serve this function—the framework of a continental journey, the echoing and development of themes from the earlier cantos. Nevertheless, we must recognise that the unification here is of the most fragile sort: the poem tends to fall apart in our minds, and to demand consideration as more than, and therefore less than, a single entity. We may feel something of the same tendency in *Gulliver's Travels*, though of course in qualities of temper that is a work of far higher standing. Unification is closer in those works where the sequence of the writing is still more or less haphazard, but where there is a general uniformity of viewpoint and texture. This is the case with the best kinds of picaresque writing, from Nashe's *The Unfortunate Traveller* to *Tristram Shandy* and *Don Juan*. *Tristram* and *Juan* could come to a stop, through the author's weariness or death: they could hardly be finished. Yet we have no difficulty in recognising them as single works. *Tristram* does not even depend on the continuous or recurrent presence of the titular hero, who is abandoned in the later books like Byron's Harold in the fourth canto of his alleged pilgrimage: rather, it is the special character of Sterne's comic view, the affection displayed towards the pretentious and sentimental puppets of his world, that keeps the fabric of the book whole. We must, in fact, recognise that unity can be achieved without reference to Aristotle's development of action through probability or necessity or to that parallel phenomenon in lyrical or meditative writing, the gradual unfolding of an idea. Nevertheless, that tighter method of unification, possible in literature and in other arts whose constructs exist within a time-sequence, in which each phase of the work will find its recognisably necessary consequent in what immediately follows, will exercise on us a special power. As we contemplate a human life in its entirety, or one of its apparently separable phases, or the history of some shared undertaking, we shall be conscious of a pattern in which each apprehensible moment is the progenitor of the next. So the work of art, the convenient means of achieving a sense of mastery over experience, will have the greater authority if it mirrors the kind of pattern that impels itself on us when we contemplate our own or another's history. Both comedy and tragedy present variations of this pattern. They begin at a point where a new departure is made, with Agamemnon's return to Argos or Benedick's return to Messina. They end with death or marriage or some other manifest completion of a phase of existence. Within those limits the march of events is steady, not subject to random variation. We accept the pattern with awe or delight, finding—despite disaster and absurdity—a sense of consolation in the completeness of the thing. Man has been destroyed or amusingly humiliated, but at least he had made his mark, has lived through his life or a phase of it and set an individual seal upon the passage of time.

Yet occasionally, especially in works of comparatively recent date, there is a unification of a subtler kind, where we have the sense of a completion of a phase while at the same time being aware of the ineluctable flux of experience of which the phase is an abstracted part. This will arise when a writer is particularly conscious, not so much of the individual personality or the isolated experience, as of the total setting within which the character or the experience exists. Peter Alexander has noted this as indicating one of the key-differences between Shakespeare's tragedies and histories. At the ends of *Hamlet* and *Macbeth* there are perfunctory indications of the future condition of Denmark and Scotland, but our attention is concentrated on the tragic hero who is dead. But at the end of each of the histories we are forcibly reminded that England's story is to continue: even after the triumph of Agincourt we are told that Henry V died young and grief fell on his people; when Richard III is vanquished, the coming of Tudor government is strongly set before us.[1] In the history plays this kind of structure impels us to regard England or Respublica, as Tillyard has put it, as providing the framework within which the dramatic characters play out their story.[2] Differently and more subtly we have the same method used in *Troilus and Cressida*: that play does not end, as *The Iliad* does, with Hector's death and funeral: after Troilus has spoken his lament for the chief man of Troy, Pandarus comes in to accost the young lover and to speak a sharp-tasting epilogue. We are not allowed to dwell on Hector's death; we are deprived of the satisfaction of finality; Troilus will go on, diverting his passion for Cressida to military undertakings, Pandarus will still seek wretchedly for the vicarious thrill. In the theatre this ending has on occasion been changed, for the kind of mood in which Shakespeare leaves us is no easy one. But there can be no question that his ending was deliberately chosen. It is like the ending of *The Wild Duck*, when Gregers Werle, awake now to the dangers of his mission, yet goes from the stage with the determination to remain the thirteenth at table, the bringer of bad luck to the easy and the blind. We can find the same kind of structure even within a brief poem. Hopkins's 'To what serves mortal beauty' is concerned, up to its last line, with an attempt to find a spiritual use for beauty of frame and face: then the poet turns briefly to 'God's better beauty, grace', which provides the larger context within which his poem is to be read. Yeats, as we saw in the opening section of this book,[3] describes in the last of his 'Meditations in Time of Civil War' his 'Phantoms of Hatred and of the Heart's Fullness and of the Coming Emptiness', and then in a concluding stanza comments on the isolation of the seer and half-wishes that he had been gifted in a way nearer to the common liking: thus the vision of the poem is set within the larger frame of the poet's life. Because to-day we lack, most of us, the sure

[1] *A Shakespeare Primer*, London, 1951, pp. 69–70.
[2] *Shakespeare's History Plays,* reprinted Harmondsworth, 1962, p. 160.
[3] See ' "The Servants Will do that for us" ', p. 11.

faith in man that is necessary to tragedy and the simple delight in the pattern of things that informs comedy, this sense of incompleteness, of the placing of the experience within the flux, is I believe especially characteristic of the literature of our time. There is a very moving example of it in the film-version of Colette's *Le Blé en Herbe*: in this instance the director departed from the pattern of the novel. The action has principally involved a boy and a girl, brought up together though not related, who have been spending a summer holiday with their family by the sea. During that time they have become lovers. At the end the moment for the return to Paris has come, and the two children— they are just more than that—have gone to look again at the sea. They are impatiently called to join the family when all the packing is done. Reluctantly they obey, and the boy, like a husband, takes the girl's hand and they climb, very much together, from the sea to the road. Here we are made to ask what the future has for these human beings who have just complicated their long relationship. Certainly one of the more pointless varieties of Shakespeare criticism is that which asks us to speculate on what happened to Rosalind and Orlando when they got back to her father's court. There the dramatist tells us precisely all that he wishes us to think of.[1] But in this film the problems of the future are forced on our attention. What we have seen was an important incident, which would have to take its place within the life-patterns of those concerned. The film-maker does not tell us what will later happen, does not know: he will perhaps deprecate the limits of his own vision. In works like this, the sense of unification coexists with a sense that the totality of experience cannot be unified, that ultimately the most skilful artificer is lost in the flux. It is a kind of provisional making that for some of us will have the greatest power.

But within the unity of a work of art we have either simplicity or complexity. Here indeed is our surest way of

[1] Cf. 'Shakespeare and the Idea of the Future', *University of Toronto Quarterly*, XXXV (April 1966), 213–28.

distinguishing between minor and major works. Longinus, in praising a passage from Euripides, says that the conception is 'rendered more forcible by the fact that the harmony is not hurried or carried as it were on rollers, but the words act as buttresses for one another and find support in the pauses, and issue finally in a well-grounded sublimity'.[1] There are, in fact, works which seem to exist only within the dimension of time, carrying us forward in a steady current of thought and feeling: these may be said to go 'on rollers'. Other works, however, seem to exist also as structures in space: we need, not to read straight on, but ever to be relating each part of the writing to every other part, noting how each word or pause modifies the significance of its contextual setting. We are forced thus to hold the whole work in our minds in a moment of time, and feel every strain of the structure. The distinction of Longinus is put in terms of the means of expression, but it is essentially the same distinction as that made by Richards in his *Principles of Literary Criticism*, between the poem in which the impulses run parallel, as in 'Break, break, break', and that in which the impulses are heterogeneous and opposed, as in the 'Nocturnall upon S. Lucie's Day'.[2] The major work of art depends on tension, on unresolved contradiction, as manifestly tragedy does. The major artist does not give us a 'message',[3] a simple feeling to take with us to the altar or the barricade. Feelings associated with religion or with revolution may be strong within his work, but simultaneously that work will exhibit a doubt, a regret, or a diametrically opposed feeling. There is a common belief that in the major works of art we shall find an expression of 'wisdom', of 'serenity', impelling us to reconcile ourselves with the nature of things. But nothing can be farther from the impact of a major work of art than the feeling of reconciliation. We

[1] *Longinus on the Sublime*, ed. W. Rhys Roberts, Cambridge, 1907, p. 147.

[2] Chapter XXXII.

[3] See above, ' "The Servants Will Do That For us" ', p. 13.

shall not, if we are wise, go to art for doctrine. Consequently the critic who is, as a man, committed to a particular doctrine may be inhibited from a just assessment, or at least will achieve it only with difficulty. He will want to reject the work that implies, as a major work will, a qualification of his own faith, or he will try to force upon a work of art an interpretation that will bring it into accord with his personal commitment. So the best critic is likely to be the man with an inner reserve of scepticism, the man unprepared to go whole-heartedly either with the crowd or with the côterie, not engaged to the limit with any cause, even with that of country or Christendom. It is a comfort for us to associate with our fellows, to share a belief, to assert the presence of a rock. It is understandable that we should seek that comfort, but we should not seek for confirmation of it in works of the highest art. For the major artist cannot give us that confirmation, even if he wishes to. In trying to gain a sense of mastery over a phase of his experience, he will—if he is to be just to that experience—give expression to its contradictions. The primary consolation he offers us is that through his work we see more clearly the tensions in our own lives. We shall be offered no solution. But perhaps we shall be fortified a little in the knowledge that in the life of the highly conscious individual there can be no settled peace, that the continuing tension challenges a welcome from us. Eliot gave eloquent expression to this at the end of *East Coker*, turning our minds away from serenity and towards the further journey, the further effort, the continuing irresolution:

Old men ought to be explorers
Here and there does not matter
We must be still and still moving
Into another intensity
For a further union, a deeper communion
Through the dark cold and the empty desolation,
The wave cry, the wind cry, the vast waters
Of the petrel and the porpoise. In my end is my beginning.

When we are content to recognise that spirit as regularly informing the major works of art, we shall begin to be qualified as critics.

Chapter 3

COMEDY IN THE
GRAND STYLE

I T was some twenty years ago that Cleanth Brooks published *The Well Wrought Urn*, one of the major prouncements of what has come to be called the 'New Criticism', particularly associated with Yale University but profoundly influential wherever English studies are followed. The volume began with an essay called 'The Language of Paradox', which modestly admitted that 'Few of us are prepared to accept the statement that the language of poetry is the language of paradox' but went on to argue cogently that poetry in its essence is a thing of paradox, a thing that always implies, for example, the idea of a re-born phoenix when it displays the urn as an image of death. Brooks's viewpoint was not novel to those who had taken as a norm the involuted manner of the early Eliot, who had seen his juxtaposition of contraries as a true poet's response to a world where neither systems nor allegiances were to be accepted so wholeheartedly as to bring peace to the mind, and who had found in the 'metaphysical' poets of the early seventeenth century a similar tension of thought, a similar paradox of manner. As early as the 1920s Richards had taught us to see the provisional character of poetic belief, the special virtue as he felt it of the poet who offered 'a music of ideas'[1] in which the concepts made up the tunes woven together in a major harmony. Perhaps the difference between Richards in the *Principles of Literary Criticism* of 1924 and Brooks in *The Well*

---

[1] The term was used with particular reference to *The Waste Land* and 'The Hollow Men' (*Principles of Literary Criticism*, London, reprinted 1934, p. 294).

Wrought Urn of 1947 was that Brooks was ultimately on the side of intellectual validity while Richards was preoccupied with the welfare of the reader's nervous system. But for both critics the poetic mode was one of paradox, of suspension; its patterns depended on a recognition that the great mass of data confronting any human mind is in essence amorphous. For both of them there were always warp and woof, ever at right angles, each denying the other's direction.

Here perhaps, to avoid a temporary misconception, I ought to put in a personal word. Though in some matters I do not go along with the New Critics—for example, in my own study of literature I am generally much more concerned with historical relationships than many of them are—in respect to the idea of paradox I am wholly with them. This is not because I want here to champion some kinds of literature against others, as will indeed become evident: it is, rather, because I cannot contemplate the facts of human experience as being other than paradoxical. Everything that happens seems to have at least two faces; no allegiance for a thinking human being can, I have suggested,[1] be ultimate, even though in order to act at all we must appear to assume that it is; no metaphysical or ethical or political system is without a touch of the absurd. As Camus saw it in *The Myth of Sisyphus*, the supreme paradox was that Sisyphus might be happy, because conscious, as he made his way down the hill to begin once again his dreadful ever-to-be-repeated ascent pushing the great stone.[2] A literature without paradox is the writing of a fugitive, a man who pretends to be blind.

Nevertheless, the kind of poetry that the New Critics have most often chosen for the demonstration of their views is not the sole repository of literary greatness. It is a poetry where contradiction is woven into the surface-fabric, into (we may say) the skin. Of course, there is plenty of such

[1] See above, 'A School of Criticism', p. 40.

[2] Albert Camus, *The Myth of Sisyphus*, tr. Justin O'Brien, London, 1955, p. 99.

43

writing. Not merely the metaphysicals and Hopkins and Yeats and Eliot and Stevens and Ransom (and that is much), but the writers of Jacobean tragedies and of many of the major novels of this century and the last: these at their best— Shakespeare and Chapman and Webster, George Eliot and Stendhal and Melville and Dostoyevsky—demand that we simultaneously give and reject an allegiance to this or that idea or code: they are masters of the deflating image, the ironic postscript, the great panorama which will make a kind of sense only if we recognise the principle of non-cohesion in our universe. 'Vengeance is mine, and I will repay' was what Tolstoy quoted on the title-page of *Anna Karenina*, but the novel leaves us brooding on the identity and the probity of the one whose words are given. This is a long distance from Hardy's pronouncement: 'The President of the Immortals (in Aeschylean phrase) had finished his sport with Tess', but it is not to be taken as a simple obverse of that: Hardy denies, Tolstoy holds us in suspense, as he does in detail after detail throughout his book. Yet, by here mentioning Hardy, I have brought in a name that does not belong in the literature of overt paradox, and once we begin to look around we are aware that the world's literature has an abundance of writers whose characteristic mode of utterance strives towards a condition of simplicity and directness. And among these writers are some of our best. We shall remember Milton's comparison of rhetoric and poetry, in which he asserted that poetry was 'less subtle and fine, but more simple, sensuous and passionate',[1] that Coleridge approved this despite his recognition that the poet's imagination worked through the reconciling, or at least simultaneous presentation, of opposites,[2] and that Matthew Arnold's insistence on the primacy of the 'Grand Style' was a continuation of the same line of thought. At this point we may look at some of Arnold's statements on the subject. Of the Greeks he says:

[1] *Milton's Prose*, ed. Malcolm W. Wallace (World's Classics), 1937, p. 154.

[2] *Coleridge's Shakespearean Criticism*, ed. Thomas Middleton Raysor, London, 1930, I, 165–6.

Not that they failed in expression, or were inattentive to it; on the contrary, they are the highest models of expression, the unapproached masters of the *grand style*. But their expression is so excellent because it is so admirably kept in its right degree of prominence; because it is so simple and so well subordinated; because it draws its force directly from the pregnancy of the matter which it conveys.[1]

The nineteenth-century writer can learn from them, he says,

how unspeakably superior is the effect of the one moral impression left by a great action treated as a whole, to the effect produced by the most striking single thought or by the happiest image. As he penetrates into the spirit of the great classical works, as he becomes aware of their intense significance, their noble simplicity, and their calm pathos, he will be convinced that it is this effect, unity and profoundness of moral impression, at which the ancient Poets aimed; that it is this which constitutes the grandeur of their works, and which makes them immortal.[2]

These passages are from the Preface to his *Poems* of 1853. In his lectures *On Translating Homer* published eight years later he declared that *The Iliad* must be the work of a single writer because it has 'a great master's genuine stamp, and that stamp is *the grand style*'.[3] He was well enough aware that he had not defined this key-term, and in the 'Last Words' added to these lectures in 1862 he offered this account of the matter:

I think it will be found that the grand style arises in poetry, *when a noble nature, poetically gifted, treats with simplicity or with severity a serious subject.*[4]

He went on to say that he thought none of these words themselves needed defining. Perhaps we shall not be so sure of that, hesitating in particular at the word 'noble'. There is,

[1] *Poetry and Criticism of Matthew Arnold*, ed. A. Dwigth Culler, Boston, 1961, pp. 206–7.

[2] *Ibid.*, p. 212.

[3] *On the Study of Celtic Literature and On Translating Homer*, New York, 1902, p. 183.

[4] *Ibid.*, p. 265.

perhaps, something of Arnold senior about it, suggesting a man who strives to present his mind as made up in the right way and to act consistently in accord with a total and unshakeable view. I am not mocking, though it may be assumed I am. There is indeed great poetry which is not only eloquent and passionate but is simple and severe and monolithic, where, in Richards' terms, the impulses seem at first sight to run parallel.[1] And we should note too the insistence in the 1853 Preface on the 'profoundness of moral impression': in further consideration of the Grand Style we shall observe how the poems in this mode powerfully represent a characteristic ethical drive of their society. Even so, one of the curiosities of literary history is that these should be the utterances of the poet of 'Dover Beach'.

Still, it can be argued, one has to look deeper for the paradox. All may be simplicity and severity on the surface, but there is a tension between surface and substratum. It is true that not every great poet implies a phoenix when he uses the word 'urn', but he may always imply chaos when he speaks of order, the difficulty of belief when he is expressing himself most piously, the dubiousness of loyalty when he is most strongly praising Augustus or Eliza and our James. The most notable essay along these lines in recent years has been William Empson's book *Milton's God*, first published in 1961. For Empson it has seemed inconceivable that Milton could think God's ways easily justifiable, yet, in the seventeenth-century context and in the tradition to which Milton belonged, to see them as unjustifiable would bring its special terror. So he set himself the task of justification, determined to find out what a skilful pleader might make of the case. 'Let us do our best for him: at least we'll see what it comes out like', he is presented as saying to himself. He would be a pleader, so he would put as favourable a view on things as could be contrived. But beneath all his skilful words, indeed deliberately hidden by them as far as it could be hidden, lay the enormous doubt, the terrifying repugnance.

[1] See above, 'A School of Criticism', p. 39.

46

It appears that Dante criticism has been touched by this too. For some readers Dante appears to be saying: 'Let us, as fully as we may, set down the cosmic scheme according to the Christian view of things; let there be a full use of decorum in the way we do it, so that a comic grotesquerie will indicate what damnation would induce in lesser men and lesser devils, a tragic kind of dignity for the presentation of a man who contemplates his punishment with sober horror or with the hope that a divine promise allows him, and—above all and with most force—the plainest words, the simplest gesture, for the ultimate wonder, the sight of God at the heart of the Rose; if we are thus objective, if we make our representation as congruent as we can with the traditional notion, then we shall be in a position to see it steadily and whole; we shall know what this thing is.' No one I think firmly believes that Dante was in grave intellectual difficulty before he started, but some readers come to feel that his task became at moments almost intolerable, as when, for example, Francesca or Ugolino entered his mind. Certainly we ourselves may flinch during the journey through hell, we may feel some resentment in purgatory, we may shrug when Beatrice in heaven finally melts into a smile. And so we may come to believe that Dante, like Milton according to Empson, was doing his best in what turned into an impossible assignment. If we accept that, we can say that the Grand Style is here paradoxical without surface-tension, that the ambivalence or paradox is on the large scale. But we have at this stage to recognise the fact that the weight of scholarly opinion is against this view, and probably rightly. We may share Empson's response to Milton's divinity and may find Dante's universe crying out for the ironic aside, however obscurely embedded. But that does not mean that Milton or Dante must have shared our view of the immensity of his task, any more than Vergil must have had doubts concerning the destiny of Rome.

For if there is this kind of paradox in such poems, it does not seem to be there in an important way. There may conceivably be moments of doubt or resentment, but if so

they do not persist: the affirmations are not merely magniloquent, they are climactic. Both *Paradise Lost* and *The Divine
Comedy* are primarily remarkable for what we may call the
steadiness of the viewpoint despite its width of range. We
may go so far with Empson as to admit that Milton needed to
proclaim his purpose of justifying God's ways, while Dante
could take the justice for granted. But with both poets the
controlling desire is to put oneself in a position to see God's
world both largely and intimately, and the method was
through the process and achievement of poetic representation. The dominant marks of the style would be severity and
simplicity; it would, in fact, be the Grand Style, the appropriate mode of expression for a representation of the cosmos
made by a man who does not actively entertain a doubt of
ultimate goodness.

Where, however, Arnold's use of the term 'Grand Style'
seems faulty is in his narrow application of it. I shall be concerned in the remainder of this essay with the suggestion that
it can be more widely applied, that—despite Arnold's
apparent denials in his *Essays in Criticism*—there is a place
in it for comedy, and that, in a sense different from any so
far used, a poet's choice of this style is the most gratuitous
and the most paradoxical act that he can perform.

First we may glance, not at comedy, but at the plays of
Racine, which constitute the supreme example of the Grand
Style in modern tragic drama. This will help to free this
style from the necessary connection with epic which hitherto,
following Arnold, I may have seemed to imply. In writing
about the tragedies of John Ford I have on occasion compared his mode of utterance with Racine's, but with the
reservation that Ford has not Racine's degree of austerity,
of paring away.[1] In Ford there are paradoxes on or near the
surface, despite the tonal simplicity: the characters engage
in intrigue, their motives are frequently dubious, they may
use violence on one another, they move through torment and

[1] *John Ford and the Drama of his Time*, London, 1957, p. 64; *John Ford*
(Writers and their Work, No. 170), London, 1964, pp. 34–5.

self-torment to an only ultimate condition of stillness. In Racine there may be intrigue too, as most obviously in *Phèdre*, but it is a datum in the story, never I think prominent in the mind of author or spectator. The extreme point in Racine's *œuvre* is reached in *Bérénice*, where Bérénice and Titus and Antiochus merely talk about the course of their suffering and its extension into their future. Titus has brought Bérénice to Rome; they love each other, but the senate and people of the empire cannot endure the thought of a foreign queen as empress, and therefore they must part; Antiochus too loves Bérénice, but there is no hope for him whatever may happen to the queen and the emperor; the final separation of all three is evident enough from the beginning of the play, and all the action consists in their coming to terms with it. This is drama without reversal, *peripeteia*, as it is without violence or intrigue. 'Dans un mois, dans un an, comment souffrirons-nous', exclaims Bérénice to the man she loves, the man who, she knows, loves her. Even in these chiselled words may be discerned a touch of paradox in that, though they suffer and will suffer through separation, they are linked together through a shared grief: although, as she says in the next line, 'so many seas may separate me from you', they can talk appropriately of 'nous'. But that is incidental: the dominant effect is that of the graven stillness on the face of ultimate grief, the refusal to behave in a common way, the deliberate stylisation of response, the freedom from the flux of life achieved by assuming the marble of the tomb. This, of course, in relation to another range of ideas, was the Greek mode in tragedy, and it has its echoes through the ages. The Greeks immobilised the passions through the use of masks and through the high authority of their lamentation.

So, too, the note of the unutterable, of an anguish presented in pure, uncomplicated terms, can appear in drama where we least expect it, where indeed the work as a whole is far from the Grand Style. George Steiner has written in *The Death of Tragedy* concerning a well-known moment in Helene Weigel's performance in *Mother Courage*:

There comes a moment in *Mutter Courage* when the soldiers carry in the dead body of Schweizerkas. They suspect that he is the son of Courage but are not quite certain. She must be forced to identify him. I saw Helene Weigel act the scene with the East Berlin ensemble, though acting is a paltry word for the marvel of her incarnation. As the body of her son was laid before her, she merely shook her head in mute denial. The soldiers compelled her to look again. Again she gave no sign of recognition, only a dead stare. As the body was carried off, Weigel looked the other way and tore her mouth wide open. The shape of the gesture was that of the screaming horse in Picasso's *Guernica*. The sound that came out was raw and terrible beyond any description I could give of it. But, in fact, there was no sound. Nothing. The sound was total silence. It was a silence which screamed and screamed through the whole theatre so that the audience lowered its head as before a gust of wind. And that scream inside the silence seemed to me to be the same as Cassandra's when she divines the reek of blood in the house of Atreus. It was the same wild cry with which the tragic imagination first marked our sense of life. The same wild and pure lament over man's inhumanity and waste of man.[1]

The grandeur of the minimal utterance, the 'pure lament' as Steiner puts it, the moment when the face becomes a mask, are what O'Neill aimed at, and at times came near to achieving, in *Mourning Becomes Electra*. And in one sense this goes beyond arbitrary abstraction, having its roots in our common experience. 'Though a quarrel in the streets is a thing to be hated, the energies displayed in it are fine; the commonest man shows a grace in his quarrel', said Keats,[2] and he was recording our occasional recognition of a momentary gesture, a short passage of unpremeditated speech, where we feel that something essential to the man concerned has been epitomised: it is the moment we remember him by when years or distance have separated us, and it is remembered too as something momentary, static, even if it involves a passage of

---

[1] London, 1961, pp. 353–4.

[2] Lord Houghton, *The Life and Letters of John Keats* (Everyman's Library), 1938, pp. 158–9.

words, a sequence of gestures. There is one mask that we make, unknowingly, for ourselves, and it is a by-way through which the Grand Style makes a direct contact with the life we live.

All literature, of course, is an attempt—in this book repeatedly referred to—to escape from the flux which is what we live in day after day. But there are two opposed ways of doing this. We may set aside various abstractions from the flux, making indeed a pattern out of them so that we have the sensation of multiple response without passivity. This is the way of Shakespeare and Donne, of Hopkins and Eliot, of Melville and Joyce. Everything the writer puts there is the result of an isolated act of perception—that is, the result of isolating a fragment of experience—but it is seen along with other such results, and the emergent totality challenges our capacity to see relationships, to sketch out a pattern which will be provisionally valid for our sense of the flux as a whole. These writers aim at inclusiveness, and the more nearly their patterns seem to stand for the whole of experience, the more highly we shall be likely to regard them. They give us, we say, a sense of 'life as it is'. Their utterance will be complex, often difficult, and in most instances open to diverse interpretations. The vast body of writings on *Hamlet* is an indication that men of talent and information have failed to agree on what its author was doing. In this extreme form we can speak of multiple paradox, where there seems no end to the discovery of the number of fragments that have been fused in the writing. Where scholars and play-directors fail with such a work is in simplification, in pretending to themselves and to us that the dramatist after all was giving a plain statement. Whatever *Hamlet* is, it is not an example of writing in the Grand Style.

That indeed works in the opposite fashion. There is again a series of acts of perception, of glimpses of the world, but there is a native congruence between them. Contrasts are incidental or constructive, never demanding resolution into a new compound; the impulses, as we have noted Richards putting it, run parallel. The poet's expression is made as

pure as possible by stripping away anything that in the
nature of things tends dangerously to adhere to it. At its
barest this mode can be found in the simple lyric as written
by Ben Jonson or Sappho. We may note in passing that the
nineteenth-century lyric often goes astray not because it is
too simple but because irrelevant material has not been got
rid of:

> Tears, idle tears, I know not what they mean,
> Tears from the depth of some divine despair
> Rise in the heart, and gather to the eyes,
> In looking on the happy Autumn-fields,
> And thinking of the days that are no more.

This is a little long-drawn-out, but the thing that may most
properly worry a reader is the word 'divine': Tennyson has
no right to bring in the idea of God when what he is concerned with is a purely human response to mutability; he
imposes our 'despair' on the notion of a divine being, which
is impertinent if it means anything at all; he imports into
his pattern a deity for which it has no room. We may contrast
the successful bareness of Ben Jonson's:

> It is not growing like a tree
> In bulk, doth make man better be;
> Or standing long an oak, three hundred year,
> To fall a log at last, dry, bald, and sere;
> A lily of a day
> Is fairer far in May,
> Although it fall and die that night,
> It was the plant and flower of light.
> In small proportions we just beauties see;
> And in short measures, life may perfect be.

We may object, on consideration, that even the lily is not
perfect, that the corruption is not merely imminent but
already there—the lily, we may remember, can be sick like
Blake's rose—but the abstraction has been made: an image
of perfect, evanescent beauty has been conveyed, as Racine
conveyed an image of total grief. And the writer choosing

this mode may go beyond the scope of lyric and may aim, as the epic writers have done, at the presentation of a cosmo-logical scheme or, like Vergil, of a nation's destiny. In between these extreme terms comes Racine's contemplation of the grief of Bérénice, the horror at herself which Phèdre feels. In all these instances the mode is that of paring down, of getting rid of 'surplusage'. The ideal was Pater's when he exclaimed: 'Surplusage! he will dread that, as the runner on his muscles.'[1] The ideal, in fact, is the fully isolated single perception, the rigorous turning away from all but the simple gesture, the brief glimpse of how men and things may be. The result is always vulnerable. After seeing or reading *Bérénice* we to-day may grow impatient. Surely, we may ask, Bérénice and Titus are foolish in their nobility: they need not separate till to-morrow morning; the interim is theirs, so let them stop this formal lamentation and quietly slip away from each other afterwards. Or we may argue that *Phèdre* is a demonstration not of human anguish but of the responsibility of non-human powers: it is an indictment of the world's government. But such after-questionings are irrelevant as we read, and surely irrelevant to the work as we should let it operate on us. What the dramatist wants is to see, and to make us see, something limited in scope and to see it with an approach to ultimate clarity.

An approach, however, only. We shall not fully get rid of the distracting detail; and perhaps this is particularly the case with literature, the art of language, because lan-guage is also the instrument of our everyday blurred res-ponses to momentary preceptions, which are never free from a marginal, inert awareness of the environment in which those responses are made. The distracting element can work even on the simplest linguistic level. When an English-speaking reader comes to the line I have quoted, 'Dans un mois, dans un an, comment souffrirons-nous', he auto-matically reacts with the thought: 'That is how we are taught to put exclamatory sentences in French, with the pronominal

---

[1] *Appreciations with an Essay on Style*, London, 1907, p. 19.

subject after the verb, while of course in English we do it differently.' It is to be doubted whether he ever quite gets rid of this part of his response, however good his knowledge of the French language may become. And in his own language he will similarly respond to the associations of words and constructions, distractedly recognising the figures of rhetoric and linking up the simplest words with the aura they have acquired through his long habituation to them. As Wordsworth realised in his Preface to the 1800 edition of the *Lyrical Ballads*, common words are embedded in the reader's experience and can be prominently used only with danger:[1] 'mother', 'love', 'suffer', 'joy', 'peace' bring with them a damaging aura unless they are qualified, or juxtaposed with contrasting terms, in a way foreign to the Grand Style. It is for this reason that poets have so often striven for a specific poetic vocabulary, which runs the opposite danger of losing contact with the perception that is being given, or is meant to be given, stable form. When Shakespeare wrote the line 'Absent thee from felicity awhile', he parted from common language, and the result has been that the line for most readers and audiences has generated merely a vague emotionalism. My own belief is that he was aiming at a special subtlety here, showing Hamlet failing to come to terms with the situation, finding a too easy felicity in death despite all his previous doubts, despite the church's rigid prohibition of the suicide that Horatio was bent on. But, if I am right, Shakespeare has been too subtle for his public during a good stretch of time.

Moreover, there are larger issues that prevent, the single act of perception from being truly isolated. We write in one Kind or another—comedy, tragedy, lyric, epic, and so on—and the Kind links what we immediately present to all other writings belonging to that Kind. So *Bérénice* belongs with *Hamlet* and *Oedipus*, and we cannot see one expression of tragic grief without thinking of other, and perhaps more

[1] *The Poetical Works of William Wordsworth*, ed. E. de Sélincourt, Oxford, second edition, 1952, II, 402.

notable, examplars: this extends the canvas, and frequently dwarfs the thing before us. Beyond Kinds, moreover, lies the poet's use of archetypes, those basic patterns of experience that provide literature with one of its frames of reference, that invoke not only folk-memory but all the previous uses to which they have been put. Of course, the idea of a Kind, the employment of archetypal reference, even the choice of a metrical pattern, are modes of abstraction: all in some measure imply a shape foreign to the flux. Nevertheless, as the reader becomes aware of each of them—and he must, to a greater or lesser degree, if they are to work fully on his response and thus justify their use—he is made to recognise links between the immediate act of perception and a host of other perceptions that he has found given shape elsewhere in literature. This will inhibit the notion of singularity, of the isolated and immune perception. No style is securely Grand, no sentence is proof against contamination from outside experience, no poem is self-subsistent. We have seen how some obstinate reflections will come into the mind even with Dante and Milton and Racine.

We may take this a stage further. The epic or divine poet may concern himself with the whole story of the cosmos or the nation he feels to be inseparable from it, as Homer and Vergil and Dante and Milton did. But the poet who, like these, aims at the expression of an isolated perception but concerns himself with a particular human situation is necessarily arbitrary in his choice. Why Thebes, or Argos? Why Bérénice and the situation in Rome? In each case we have a personal story, the situation in a particular *polis*. The aim is something as depersonalised as the tragic mask, but the personal and social details are stuck on to it and bring it into our relationship with an environment which complicates our response. Racine wants us to think of Titus and Bérénice, but the presence of Rome invokes the whole story of the empire: not only that, but the particular figures of Titus and Bérénice make us think of this person and that, not of the essence of separation and its grief. Of course, Coleridge said that the Poet's imagination worked through the fusion of the general with the concrete,[1] but it may be doubted whether the fusion is ever complete. The particular imagined figure—whether Racine's Bérénice or Shakespeare's Cleopatra or Crashaw's St. Teresa—will persist in forcing her presence on us. We cannot get to the mask itself even in Greek tragedy, for behind it we are always aware of the particular problems, the particular responses, of a Clytemnestra or an Antigone. *Paradise Lost* itself is in some measure a story of particular people: Satan, God the Son, Adam, Eve, the talkative angel. All are in some measure psychological studies, and we are bound to ask how things would have been if any one of the persons had been a little different. They behave in such-and-such a way because they react to circumstances characteristically. And we feel this in Dante too, as he tells us why a particular man is in a particular circle of hell, a particular place in purgatory or heaven. Always there is a sense of the arbitrary: why is this one chosen for portrayal and not that? When we come to the seventh circle of hell, we feel something of surprise at finding Brutus and Cassius so prominent, whatever allowance we may make for Dante's ideas about the empire. Surely, we think, there was a case for putting one or two others along with them? Nor can we convince ourselves that we are being wholly unhistorical in finding him arbitrary in his choice.

Indeed there is something arbitrary in any act of perception. Why should we look here and not there? We choose to contemplate this rather than that, but it would be difficult to argue for an essential rightness or inevitability in the choice. It is like the arbitrariness of falling in love (a quintessential act of perception), where we cannot logically demonstrate why it is with this person rather than that. And if choice does not truly enter into such things, if we are the slaves of circumstance in what at any moment we appear to choose to isolate from the flux, it still seems arbitrary as we contemplate ourselves in the act of perceiving or contemplate others in the act of giving external shape to a perception. We may say that

---

[1] *Biographia Literaria* (Everyman's Library), 1910, p. 166.

Vergil gave expression to a Roman myth because he was a Roman living in the time of Augustus, that Dante and Milton constructed their topographical and historical accounts of the universe because these particular accounts were appropriate to their times and places. But surely, even with the firmest of faiths, there is a minimal sense that the vision achieved and offered is parochial? If 'to make an act of faith' means anything, it must imply that in making it one chooses out of a series of possibilities, one declares an allegiance, one says 'I shall see it in this way'. And to declare an allegiance implies the recognition of other possible allegiances. Only the second-rate intelligence fails to see an act of (as we may term it) choice in declaring for the established doctrine of his period. The act of choice is only more obviously, not more essentially, a choice when it goes against the current.

That does not mean that Dante or Milton or Vergil counterpoints his expression of faith with an expression of dubiety. Rather the dubiety is the framework within which the act of perception is made: it constitutes the sense of the arbitrary which is involved in all choice. The simpler the mode of utterance, moreover, the more obvious the arbitrariness. We get far less of a sense of arbitrariness in the kind of writing we have associated with Shakespeare and Donne: their commitment remains obscure; they are concerned with a simultaneous recognition of widely diverse particulars; the pattern offered aims at bringing as much as possible of the flux into a momentary coherence. Of course, they do abstract, but, through the use of overt paradox, through the deliberate collision of word and word, they give us the impression of remaining within, yet in some degree mastering, the flux itself.

Choice has something brutal in it. It is partial, in both senses of that word. To perform any definitive action is to disregard a myriad of warning precedents. To choose one viewpoint—that is, to make any act of perception—is to turn away from a host of potentialities. In our lives as individuals and citizens, we correct and punish and acquiesce in

57

correction and punishment. In doing so we set ourselves up as arbitrators, we interfere with others, we assert rights and powers over them. We choose to make ourselves and others go this way rather than that. It is in the interest of abstract right, we say, or it is for society's good. That is an act of faith, and no society can function without it. In truth all *tyrannoi* are tyrants, all societies are aggressors against the individual, all individual action is a partial rejection of the world of our total being. In these circumstances it is not surprising that we are attracted by the notion of stoic withdrawal, by the task of making something apart from ourselves and our fellows, something that exists outside society and is not an attempt to communicate with it:[1] to be involved with others is to use our power against them, to feel their power exerted on us; to take part in a communal act is to become tainted by communal arrogance. We may remember how in André Cayatte's film *Nous sommes tous les Assassins* it was painfully argued that a priest who ministered to a man under sentence of death was acquiescing in the social killing of a man, was assisting society in one of its moments of barbarity. So it may seem better to stay apart and, as the stoic put it, to be king over oneself. But of course we cannot stay apart, except by suicide, and even that is an act involving arbitrary choice: we are both individuals and members one of another, and to choose either the ultimate isolation or the community's corruption is to go against an essential part of ourselves. Every man wants to be a poet; no poet is a poet all the time. Nevertheless, the more fully and unreservedly we commit ourselves, the more extreme and manifest is the act of rejection.

This becomes more manifest, of course, from outside the immediate context within which the choice is made. We are more conscious of what Dante rejected than any fourteenth-century Italian could have been. Our own choices will be seen as perhaps equally arbitrary in other places and times. Literature in the Grand Style, however, carries its choice

---

[1] See above, '"The Servants Will Do That For Us"', pp. 13–14.

as a banner. It will see things this way and not that, it chooses these people and this action and declares they represent an ultimate. We have seen they do not, but such an awareness is disregarded and thus defied. Severity and simplicity are possible because this way of thinking, this way of writing, are seen as 'noble'. And here we may, at last, find room for comedy. If a writer cuts away non-essentials, refuses to entertain dubiety within the frame of his work, concentrates on a single point of view, exposes human folly without mercy and exalts the human wit in its manipulation of event, he gives us an analogue to Racine's tragic drama. As Bérénice aspires to the condition of the masked figure of Greek tragedy, so the comic personage in the Grand Style aspires also to the corresponding condition in Greek comedy. But, let there be no mistake about it, this will be a cruel comedy. To match an example of tragedy in the Grand Style, we may take an example from classic French comedy, Marivaux's *Le Triomphe de l'Amour*. It tells how Léonide, Princess of Sparta, fell in love with Agis, the rightful heir to Sparta's crown, and disguised herself as a man in order to find and woo him in the retreat where he lodged with the philosopher Hermocrate and his sister Léontine. So that she may persuade brother and sister to let her stay awhile with them, she pretends in her male disguise to love Léontine and, admitting she is a woman, to love Hermocrate. Brother and sister respond, after some substantial resistance, by falling in love with her. She has a scant and passing word of pity for their situation, but the stress is wholly on her ingenuity and on the triumph which love, operating through her, has over Hermocrate and Léontine and Agis. Now Marivaux has not, I think, a great felicity with words. He is even, perhaps, a little loquacious rather than eloquent. Yet the web of his intrigue is finely meshed: there is nothing to be spared in the action, and there is almost no sentiment except that which belongs to the characters alone and is thus part of a pattern made exemplarily free from the world of experience out of which it has come. The play is so calmly and elegantly brutal as to possess a special air of authority. It makes free with human beings as a convinced aristocrat might, and in that sense at least it can claim 'nobility'.

'The angels all are Tories', said Byron in *The Vision of Judgment*. So, too, perhaps, are the writers of comedy in the Grand Style. Aristophanes is the eldest among them, and his followers have not disowned him. They are always conscious of society and its ways, while treating that society with manifest contempt, and their special delight is to show men and women manipulating circumstances within a world that is given to them. They are without repining and without much mercy: they see the primal joke, but ensure, in their writings, that it is made at the expense of others. When a comedy expresses doubt about the way things are, or takes a double view of its protagonist, or grows bitter or nostalgic, it cannot belong with the comedies in the Grand Style. Thus Shakespeare and Chekhov are not, in their most characteristic comic writings, authors of this kind. Perhaps Shakespeare came nearest to the Grand Style in *Much Ado about Nothing*: only a little less concern with Hero's and Beatrice's plight would have made this as cruel a piece of writing as *Le Triomphe de l'Amour*. Elsewhere in comedy his manner is as many-faceted, as counterpointed and paradoxical, as it is in his tragedies and histories. Of all English writers, Congreve achieves the Grand Style most brilliantly and perhaps Wycherley in *The Country Wife* most purely. To-day we often sentimentalise Congreve, making a little too much of his concern with proper relations in marriage: there is, it is true, something of a sententious strain in him, in *The Double Dealer* and *The Way of the World*,[1] but at least in *Love for Love* he is singularly free and arbitrary in his exhibition of folly and of ingenuity in coping with folly. When comedy is noticeably free from the sententious, when it makes its action spare and neat, when its words shine and do not reverberate, it

---

[1] For an attempt to explore Congreve's relation to the recently developing sententious style, and to show how this both qualified and enriched his comic writing. 'Congreve and the Century's End', see below, pp. 172–96.

becomes the complement of Racinian tragedy. It is just as arbitrary, as restricted. Its act of faith is a strangely limited one, one indeed that no man could stay with long. It is small even in its cruelty, but the manner of it is Grand.

And the supreme exemplar is probably neither Greek nor French nor English, but is an Italian opera with music by an Austrian and a libretto based on a French play—*The Marriage of Figaro* in the reworking of Mozart and Da Ponte. The cruelty is here so authoritative as to compel inattention from an audience that hears the music and watches the intricate pattern of event. All works out with everyone acquiescent, all is restored to harmony, in every sense, after moments of repining and frustrated efforts to undo an ordained web. Only in retrospect, perhaps, does the spectator come to see that men are here intricately subject to each other's compulsion and must accept this with as good a grace as song can teach them.

Perhaps, too, there is such a thing as the Grand Style in reverse. I have been led to hazard a guess at this through the experience of seeing Harold Pinter's *The Homecoming*. Here we have the most brutal of language, the most brutal of situations: the only grandeur comes from the simplicity of the figures, of the events and words. Never in drama has there been a family more totally bent on reciprocal hurt and therefore more vulnerable. There is, I think, no ambivalence here. What is demonstrated is an intricate pauperism: the son who appears to have escaped to a professorship of philosophy in the New World is a fugitive from everyday pressures, finding an insecure refuge with his swimming-pool, his agreeable house, his desiccated teaching, his new family-façade; the woman he has married is a fugitive too, from a shabby servitude to dubious 'modelling'; the men of his family, to whom for a short while he returns, range in their ways from resentful humility (the car-driver) to moronic aspiration (the not-so-good boxer) to ineffective bullying (the father) to second-rate, impotent cleverness (the pimp). We see how the paupers in England, ready to exploit the émigrée wife, are easily subjected to her. We see how the

intellectual grubbiness of the émigré son's existence is exposed. There is a strong suggestion, too, of total impotence
among all the men: no child is securely assigned to a father,
no man is sure of living other than vicariously. As in the way
of comedy in the Grand Style, there is a minimal action, a
mere shifting of the pieces on the chess-board, with no
moment of check-mate, no disturbing vision of new truth, no
hint of double meaning or an alternative way of seeing.
Racine said, in effect: 'This is how you might face grief if you
had my characters' simple nobility.' Marivaux and Congreve
exhibit elegant and cruel conduct. Pinter in his latest full-
length play shows the cruelty without the elegance, presents
indeed the ultimate extreme from elegance. But he does it
with the economy, the simple drive, of his predecessors in
this mode of comedy. It is a rash man who laughs at it: we
may see the play as a special variant on an established
pattern, a novel contribution to a painfully Grand tradition.
Its special horror is that it does not raise doubts: it merely
asserts a negation. It gives us 'nobility' only in its refusal to
modify its statement.

The Grand Style—in comedy as in tragedy and epic and
lyric—forbids rebellion and dubiety. It selects a point of
view, adheres firmly to it, and gives it expression in arbitrarily chosen ways. For the term 'Grand Style' to be justified,
the thing achieved must compel our attention, must give us
a sense of confrontation with an object or obstacle to be
reckoned with. It can be frightening, as any act of faith is
capable of being. When a man chooses to write in this way,
he is rejecting much of experience in the interest of coherence, of the sense of mastery. In choosing the way of simplicity, he is abstracting most finely, he is indeed man at his
most arrogant, perhaps at his most desperate. The most
chiselled art is that which has been made because life has
pressed most strongly against the artist and the society he
represents. The power of hell had to be felt with especial
strength before *The Divine Comedy* could be written; *Bérénice*
exalted love in a society where its importance was so minimal
that its rehabilitation was strained for; *Le Triomphe de l'Amour*

was an assertion of aristocratic arrogance in eighteenth-century France; *Love for Love* declared gentility's right to practise effrontery in a world where trade gathered strength; *The Homecoming* could not be written except in a time when the desire for acquisition had become so palpably impertinent. The athlete gets rid of surplusage so that he does not need to look over his shoulder; in Harold Pinter's play the dramatist shows us people who concentrate their minds because they dare not look back. It will not do to hazard a guess about the reasons why *The Iliad* has the Grand Style too: we cannot adequately know what pressures lay on the society of Homer or of the poet of *Beowulf*. But in later centuries we may be permitted to think that the cult of severity, of simplicity, of aristocratic rejection (and there are aristocrats of a sort in perhaps every community), is a mode of implying, and countering, the terror or dubiety which the writer feels in relation to the human condition as a whole. His utterance is therefore selective. His eloquence is the counterpart of a major silence. His decision to speak is itself paradoxical.

Chapter 4

WHEN WRITING BECOMES ABSURD

THE scene is Asia, near Babylon, and the conqueror
Tamburlaine enters in his chariot. Instead of by horses,
the chariot is drawn by the captive kings of Trebizon and
Soria, *'with bits in their mouths'*. Tamburlaine holds the *'reins
in his left hand, and in his right hand a whip with which he scourgeth
them'*. His progress is necessarily slow, and his followers
keep pace with him on foot. Led as prisoners are the kings
of Natolia and Jerusalem, who will take the place of the
other kings when exhaustion overcomes them. But first
Tamburlaine speaks, urging his human steeds to greater
effort, reproaching them for their inefficiency and for
their insensibility to the honour he is doing them:

> Holla, ye pampered jades of Asia!
> What, can ye draw but twenty miles a day,
> And have so proud a chariot at your heels,
> And such a coachman as great Tamburlaine,
> But from Asphaltis, where I conquered you,
> To Byron here, where thus I honour you?
> (Part II, IV. iii. 1–6)[1]

The stage-picture and the words spoken incorporate an
extreme brutality, an assertion of a conqueror's power, a
demonstration of that conqueror's need to assert his power,
a manifest folly in imposing on men a task which horses
could undertake with more success, and an extravagance
of manner which invites mockery. And for some decades

[1] *Tamburlaine the Great*, ed. U. M. Ellis-Fermor, revised edition, London,
1951.

64

there was a good deal of mockery of this scene. A dramatist could be sure of a laugh if he put into a character's mouth the phrase 'pampered jades', as Shakespeare did in the Second Part of *Henry IV*, as Jonson, Chapman and Marston did in their comedy of London life *Eastward Ho*. In the mouth of a drunkard or a windbag the words of Tamburlaine became simply comic. But as Marlowe himself used them, their effect is more complicated. They are comic with him, too, but in a painful way. When Shakespeare's Pistol utters them, there is an obvious inappropriateness of words to speaker. When Marlowe's Tamburlaine utters them, the speaker is the most appropriate one that can be imagined, but he is still fantastically inappropriate. The residual gap, between what even this man is and what he says and does, is enormous, terrifying, absurd.

From that I want to turn to Crashaw's 'A Hymn to the Name and Honor of the Admirable Saint Teresa'. We shall recall how he shows St. Teresa in her childhood, filled with the knowledge and the love of God and determined to share that knowledge and that love with the unregenerate Moors. If they will not listen, but reward her with death, her martyrdom will be an act of faith, an image of the Atonement, a source from which a blessing may come to her murderers. The poet here uses images of trade and of the sowing of seed. To purchase her crown of martyrdom the saint will offer 'her dearest Breath, With CHRIST's Name in it'—that is, both the words of the gospel which she will speak to them and the ultimate breath of her body as she dies invoking Christ. The image of sowing is an ancient one, of course, but here the seed is Christ's blood or Teresa's. And then Crashaw forsakes 'metaphysical' imagery, cultivating a simple rhetoric as he presents the child's farewell to her home. Certain bare words are emphasised and amplified through the use of capital letters: the 'FARE-WELL' the child offers to the world (the emphasis suggesting a wish for universal salvation), the 'MOTHER' and 'FATHER' (becoming not only general images of shelter but echoes of Christ's leaving Nazareth to take up his ministry), the

65

'MOORS' (whom one way or another she will conquer, who will give her the crown she seeks), the 'MARTYRDOM' that is her prize. In this short passage Crashaw shows us two different modes of utterance—first through the use of relevant images and then through the use of basic and reverberating symbols:

> Sh'el to the Moores; And trade with them,
> For this vnualued Diadem.
> Sh'el offer them her dearest Breath,
> With CHRIST'S Name in't, in change for death.
> Sh'el bargain with them; & will giue
> Them GOD; teach them how to liue
> In him: or, if they this deny,
> For him she'l teach them how to DY.
> So shall she leaue amongst them sown
> Her LORD'S Blood; or at lest her own.
> FAREWEL then, all the world! Adieu.
> TERESA is no more for you.
> Farewell, all pleasures, sports, & ioyes,
> (Neuer till now esteemed toyes)
> Farewell what euer deare may bee,
> MOTHER'S armes or FATHER'S knee
> Farewell house, & farewell home!
> SHE'S for the MOORES, & MARTYRDOM.[1]

Now, the situation can be seen as a pathetic or gently amusing one. Here is a small child, enthusiastically pious, avid for a glory and a suffering that she cannot understand, setting out to offer Christ to the Moors as a boy might run away from home to go to sea or run away from a boarding-school to go home. And, like such a boy, Teresa is brought home and preserved for the 'milder MARTYRDOM' of her later years. Indeed, Crashaw does not neglect the pathos of it: we shall have noticed the line 'MOTHER'S armes or FATHER'S knee'; he refers persistently to how young she is, a mere six years of age; he employs such adjectives as 'gentle' and 'sweet' to underline her physical frailty and infant charm.

[1] *The Poems English Latin and Greek of Richard Crashaw*, ed. L. C. Martin, Oxford, 1957, p. 318.

But the essence of the poem is in Teresa's strength, in the magnitude of her aspiration, the glory of her achievement. Like a baroque painter or sculptor, he rejoices in contradiction, in the proleptic existence of the great saint within the inarticulate child. In the passage quoted he works towards a directness and bareness of expression, embodying a notion of the gigantic in the jejune words and the easily sentimental references. It is doubtful whether he could have done this without having the basic Christian paradoxes to support him—the image of an omnipotent god as a baby at the breast, the image of that god as a sacrificial victim, the image of man's reconciliation to his god being achieved through man's murder of that god.

In passing we may note that Wordsworth's exalting of the child in his 'Immortality' Ode, as the 'mighty prophet' and 'seer blest', as the partaker in a vision and an understanding that will later be impaired, seems merely rhetorical and wishful when we compare his poem with Crashaw's: Wordsworth is toying with, and vaguely Christianising, a Platonic fancy; Crashaw, along with other baroque artists, draws life from a profoundly absorbed tradition and in return gives that tradition fresh nourishment. Yet Crashaw as much as Wordsworth writes regardless of commonsense: he defies us merely to smile at the runaway girl, he insists upon her grandeur; he uses the language of pathos, but demands homage; he implies that only the absurd is worthy of our belief and our trust.

Crashaw's poetry was admired even in Protestant England, but already in 1652—when the 'St. Teresa' Hymn appeared in his volume *Carmen Deo Nostro*—there were new ways of writing evident, and soon his defiance of commonsense became unfashionable, odd, extravagant, not quite 'English'. In associating him with Marlowe, I have made a junction which may itself appear extravagant: on the one side the rebel, the innovator, the man who in *Tamburlaine* is enraptured by a purely human glory whose comic futility he simultaneously contemplates; on the other side the man who rediscovers and revivifies a tradition, who finds a crown in self-sacrifice, and some form of martyrdom a necessity. But if 'Holla, ye pampered jades of Asia' can provoke laughter in the playhouse, and 'SHE'S for the MOORES, & MARTYRDOM' an uneasy look on the reader's face, it is because both poets are concerned with a vision of the world in which commonsense standards have ceased to apply.

And with these poets in this regard I want to associate certain modern dramatists who have taken it on themselves to expose the illogicality by which men live. We can think, for example, of Samuel Beckett in *Waiting for Godot* showing us Vladimir and Estragon as they think up devices for passing the time which they have not the effrontery to bring to a stop. They try to convince themselves that they live by a hope which they know is empty. All that the hope does is to provide them with a reason for the avoidance of planning; and that too would be a delusion if they attempted it. In the same play, when Lucky at last speaks, his words form a burlesque of an academic lecture, a lecture given by a man with a store of information which he cannot assemble into coherence. It is the embodiment of a nightmare which every lecturer has, an image of the final lecture he will give. Or we can think of Eugène Ionesco and N.F. Simpson delighting to expose the non-sequiturs, the tangential associations, which fill the stream of everyday speech, and to demonstrate the acts of violence, the blunders and the anticlimaxes that attend on human planning; or of Harold Pinter's preoccupation with mystery, the presence and dominance of forces which the human being can hardly recognise, can in no way explain. Stanley in *The Birthday Party* is pursued, persecuted, and finally abducted by the representatives of a power which we are given little help in identifying. The tramp in *The Caretaker*, with all his petty greed and his would-be smart behaviour, is utterly lost in a world where the two contrasted brothers are elemental beings.[1] In such

[1] For a fuller comment on the separateness of the brothers from the tramp, see 'Two Romantics: Arnold Wesker and Harold Pinter', *Contemporary Theatre* (Stratford-upon-Avon Studies), London, 1962, pp. 11–31.

a world—which, it is implied, we live in too—it does not help to be smart or to go down to Sidcup to get one's papers.

These dramatists are working in a time when, for themselves and for much of their audience, traditional ideas concerning the nature of the cosmos have ceased to exert an appreciable power. That is a datum for the writers: what they are immediately concerned with, however, is human irrationality. If an old god has been dethroned, man has not usurped his place. Man does not, in his moment-to-moment existence, properly exercise his 'god-like reason', and even if he did it would not help him. For the most part the plays are not morally nihilistic: Beckett and Pinter, in particular, display a hierarchy of values in their implied comments on human behaviour. Even Ionesco in *Rhinoceros* suggests a virtue in staying loyal to one's own human nature. But the presence of such moral principles is part of the mystery the plays present: the fact that man has standards of judgment, the fact that we can see these standards as in some measure valid, the fact that there seems to be a point in living as a man—these things exist in a metaphysical vacuum. They are themselves illogical, along with most human thinking and behaviour.

The plays invite laughter, first because they seem to depart so widely from the commonsense we forgetfully believe we live by, and then because we see that the picture offered is a faithful image of an absurd actuality.[1]

In these varied instances so far referred to, it is obvious we are dealing with something very different from the literature that takes on an appearance of absurdity for subsequent generations when the particular circumstances of composition no longer operate. To-day we may go to laugh at a revival of a Victorian melodrama, mocking the hero and his bravery, the heroine and her endangered virtue, the villain and his puny menace. Similarly we can be amused at the devices of popular films made in the silent period

---

[1] Above, in 'Comedy in the Grand Style', pp. 62, I have suggested that Pinter's later play *The Homecoming* defies the spectator to laugh.

of cinematography. And so it was with the people who
laughed at Henry Mackenzie's lachrymose novel *The Man
of Feeling* only a generation later than the time when its
characters' frequent tears were freely shared. Though it
may have been Wilde who said that one would have to have
a heart of stone to read about the death of Little Nell without laughing, there are stretches of *The Picture of Dorian
Gray* which we in our turn may find unintentionally risible.
But in all these cases it is the less-than-first-rate we are
dealing with, and the laughter arises because a no longer
viable technique allows the commonplaceness, and the
falsity, of the utterance to stand naked before us. Many a
popular novel or film to-day is just as commonplace, just
as false, as *The Man of Feeling* or *East Lynne*, but it is presented
in a fashion that we are accustomed to accept: more discerning readers and spectators may be irritated by such things,
but widespread laughter can come only when the current
mode of utterance has changed. In contrast, Tamburlaine's
cry to his human steeds was simultaneously magnificent
and absurd to his own generation; Crashaw's 'St. Teresa'
Hymn was a deliberate flouting of commonsense; and
Beckett and the other modern playwrights I have mentioned
aim directly at our recognition that their absurd world is
ours.

Certainly there are times when the sense of the paradox,
the basic and absurd contradiction, seems particularly
strong. As Walter Kaiser has pointed out in his book
Praisers of Folly, Erasmus—hinting at the character of true
wisdom through the mouth of Folly speaking in praise of
herself, hinting thus that wisdom must subsume folly
in order to become wisdom—is a son of the Renaissance,
while for the Middle Ages jest and earnest stood in a relation
of simple antithesis.[1] Erasmus invites us to think Folly's
utterance absurd, and then invites us to recognise its
absurd truth. So, later in the sixteenth century, Marlowe's
great conqueror Tamburlaine is both absurd and magni-

[1] Cambridge, Mass., 1963, p. 60.

ficent, both the man supreme among men and the man reduced to having his chariot drawn by inefficient kings in order that he may persuade others and himself that his power transcends the human. If we merely laugh at Tamburlaine, we are blind to Marlowe's play as a whole: among the men here presented, Tamburlaine is indeed supreme. His son Calyphas can mock him, and in a sense quite rightly. The virtuous Olympia can suggest a scale of values which Tamburlaine knows nothing of. Zenocrate can express Everywoman's fears for the risks that her husband runs. But Tamburlaine in his daring, in his success during the time allowed him, demands our wonder. It is only in the face of the ultimate human vulnerability that he is ripe for an absurd destruction, and before that an absurd exposure. In the following century Crashaw found himself in a Europe now largely, though not predominantly, Protestant, and found too in the artists of the Counter-Reformation a readiness to assert, and glory in, the illogical. His words, especially because they are written in English, take on an appearance of defiance: they emphasise the supreme quality of saintliness in an infant, thus proclaiming a kind of human magnificence that the reformers questioned. Nevertheless, Crashaw recognised that the paradox was extreme, that only faith could see the major saint in the six-year-old child, could see 'SHE'S for the MOORES, & MARTYRDOM' as a statement triumphantly absurd. And in our own day the absurdity perhaps chiefly lies in the residual act of faith, which I have referred to in relation to the persistent moral standards of the dramatists. But the Renaissance, the Counter-Reformation, the Age of the Bomb—these are surely exceptional periods, it may be argued, to be constrasted with the times when literary utterance has referred to commonsense as a norm and has demanded that men live as their shared, civilised standards dictate. Yet it is worth remembering that Pope could not refrain from tears when he read aloud the ending of *The Dunciad*, with its indication of a coming world where the light of reason was totally obfuscated. It is worth remem-

71

bering, too, that Pope could present, without a smile, a picture of the long procession of hearses that he hoped might be required through the deaths of members of his Unfortunate Lady's family:

> Thus, if eternal justice rules the ball,
> Thus shall your wives, and thus your children fall:
> On all the line a sudden vengeance waits,
> And frequent herses shall besiege your gates.
> There passengers shall stand, and pointing say,
> (While the long fun'rals blacken all the way)
> Lo these were they, whose souls the Furies steel'd,
> And curs'd with hearts unknowing how to yield.
> Thus unlamented pass the proud away,
> The gaze of fools, and pageant of a day![1]

At first it may seem remarkably coincidental that so many apparently natural deaths should occur in one family in so brief a time. But then we shall recognise that ultimately the long procession must assemble itself, that the death of the unfortunate lady will necessarily have its many parallels in the deaths of the relations who repudiated her. Ultimately, indeed, 'frequent herses shall besiege your gates', and the pride of family will seem empty when the last heir joins his forefathers. By simply speeding up the sequence—and that is legitimate enough, *sub specie aeternitatis*—Pope has claimed a fantastic vengeance for his lady's unregarded demise. And yet, we know, nothing is surer than such vengeance, not by a special divine intervention but by the operation of natural processes of decay. The sequence of the funerals first makes the poet's threat ludicrous, and is then recognised as plain matter of fact. Yet absurd fact, too, as we see a Georgian mansion with a hearse never absent from the gate; and the basically absurd element is not the procession of hearses but the mansion. Pope may be seen as exceptional for the keenness of his eye and the

---

[1] 'Elegy to the Memory of an Unfortunate Lady', ll. 35–44 (*The Poems of Alexander Pope*, ed. John Butt, London, 1963, p. 263).

frankness of his pen, as he presents the dying Villiers'
George and Garter dangling from the bed in 'the worst
inn's worst room', and proclaims man as 'a Being darkly
wise, and rudely great'; and we may note that, just as he
increased the normal frequency of funerals in his 'Elegy
to the Memory of an Unfortunate Lady', so he exaggerated
the squalor within which the historical Villiers came to
his end. That, in essence, was not a falsification: the end,
as he saw it, was necessarily, because universally, squalid,
and Pope simply laid on the colours thick so that the most
blinkered reader could not turn his eyes away.

If we turn elsewhere in the eighteenth century, we may
come upon a vision terrifying and terrifyingly comic as the
poet or novelist contemplates external nature or the world
of man. It was a time when the sanity of poets seems to have
been most precarious or when poetry fell most notably
into the hands of those men who suffered in their minds.
Here is Christopher Smart in his *Song to David* imagining
a world writhing in adoration of its maker:

> For ADORATION seasons change,
> And order, truth, and beauty range,
>   Adjust, attract, and fill:
> The grass the polyanthus cheques;
> And polish'd porphyry reflects,
>   By the descending rill.
>
> Rich almonds colour to the prime
> For ADORATION; tendrils climb,
>   And fruit-trees pledge their gems;
> And ivis with her gorgeous vest
> Builds for her eggs her cunning nest,
>   And bell-flowers bow their stems.
>
> With vinous syrup cedars spout;
> From rocks pure honey gushing out,
>   For ADORATION springs;
> All scenes of painting crowd the map
> Of nature: to the mermaid's pap
>   The scalèd infant clings.

73

The spotted ounce and playsome cubs
Run rustling 'mongst the flowering shrubs,
    And lizards feed the moss;
For ADORATION beasts embark,
While waves upholding halcyon's ark
    No longer roar and toss.[1]

Here Smart sees all time as one, making the entry into Noah's ark a permanent phenomenon, and juxtaposes the observable beasts and plants with the mermaid that man's fancy has contrived. This, after all, is man's Nature, the world he has seen himself living in. The passionately devout Smart sees that Nature as God's—for man, its part-creator, is God's creature—and sees it too as offering homage to the ultimate maker in its being and movement. It is a universe to drown the senses in, where the ordinary demarcations of space and time, of observation and fancy, no longer exist.

At another extreme, yet with a similarly oppressive fecundity, Laurence Sterne found always a world defying, yet simultaneously demanding, credulity. Whether it was the speculations of theologians on the goodness of God and the number of the damned, or the relation of writing to literary criticism, or the human response to grief or concupiscence, Sterne saw the puppets jerking at their strings' ends—likeable enough, but without self-mastery and becoming more absurd as they pretended to dignity. Here is a scene in the Shandy kitchen, where Corporal Trim harangues his fellow-servants on the subject of death, the removal of Tristram's brother Bobby having just been reported to them from above:

To us, Jonathan, who know not what want or care is—who live here in the service of two of the best of masters—(bating in my own case his majesty King William the Third, whom I had the honour to serve both in Ireland and Flanders)—I own it, that from Whitsuntide to within three weeks of Christmas,— 'tis not long—'tis like nothing;—but to those, Jonathan, who

[1] *A Song to David*, ed. J. B. Broadbent, London, 1960, pp. 20–1.

know what death is, and what havoc and destruction he can make, before a man can well wheel about—'tis like a whole age.—O Jonathan! 'twould make a good-natured man's heart bleed, to consider, continued the Corporal (standing perpendicularly), how low many a brave and upright fellow has been laid since that time!—And trust me, Susy, added the Corporal, turning to Susannah, whose eyes were swimming in water,—before that time comes round again,—many a bright eye will be dim.—Susannah placed it to the right side of the page—she wept—but she court'sied too.—Are we not, continued Trim, looking still at Susannah—are we not like a flower of the field—a tear of pride stole in betwixt every two tears of humiliation—else no tongue could have described Susannah's affliction—is not all flesh grass?—'Tis clay,—'tis dirt.—They all looked directly at the scullion,—the scullion had just been scouring a fish-kettle.— It was not fair.—

—What is the finest face that ever man looked at!—I could hear Trim talk so for ever, cried Susannah,—what is it! (Susannah laid her hand upon Trim's shoulder)—but corruption?—Susannah took it off.

(*Tristram Shandy*, Book V, Chapter IX)

Sterne's staccato prose, with sentence or phrase linked to its successor by a dash, as if no man has the stamina to escape from the flux of things for more than a few words at a time, increases the sense of a puppet's jerkiness—as do the erect posture of Trim during his oration, and the rapid movements of Susy's hand.

Once called 'The Age of Prose and Reason', the eighteenth century was a time when a man felt it necessary to abandon the Happy Valley and when one could tolerate the utterances of Pangloss only if one concentrated attention on one's garden. Comedy, our text-books tell us, is society's mockery at deviation. But the profounder comedy—of Sterne, of Ben Jonson, of Voltaire, of Aristophanes—is the product of man's incredulity as he contemplates himself. It is not the simple assertion of a norm, nor is it the safe mockery of that norm. In Isak Dinesen's *Anecdotes of Destiny* we read: 'Elishama began to realise the value of what is named a comedy, in which a man may at last

75

speak the truth.'[1] This is to claim for comedy an inclusiveness which, when comedy stands alone, is more than doubtful. But it does suggest the radical character of the more ruthless comedy, which is not content with sending us vicariously to bed.

The nineteenth century, too, has its pictures of man realising his position in an incomprehensible universe. Whether we think of James Thomson's *The City of Dreadful Night*, where the Industrial Revolution's urban monster becomes an image for a world of private nightmare, or of Dostoyevsky's *The House of the Dead*, where a humanly contrived imprisonment represents the chains that every man wears; whether we think of the grotesques of Dickens, or of the brittle pretty women like Thackeray's Becky or Trollope's Lizzie Eustace; whether it is Captain Ahab that is in our minds, pursuing the monstrous whale through the more monstrous seas, or Hawthorne's Hester Prynne damming the tide of life in her silence and deprivation— the closer we look at such writings, the more aware we become that terror and laughter are fused within them. This was, after all, the century of Edmund Gosse's childhood, described with reticent authority in his *Father and Son*. And earlier than that, in the middle of the century, Charlotte Brontë could show her bewilderment at her sister's novel in her preface to the 1850 edition of *Wuthering Heights*. She must withhold moral approval from the book and its chief characters, and can aver that Emily Brontë, if she had lived, would have given us writings with 'a mellower ripeness and sunnier bloom'. But she is honest and brave enough to recognise the limited power of the artist's volition. She stands awed but still proud as she contemplates her sister's demonic possession:

> Whether it is right or advisable to create beings like Heathcliff, I do not know: I scarcely think it is. But this I know: the writer who possesses the creative gift owns something of which he is not always master—something that, at times, strangely wills and

[1] London, 1958, p. 183.

works for itself. He may lay down rules and devise principles, and to rules and principles it will perhaps for years lie in subjection; and then, haply without any warning of revolt, there comes a time when it will no longer consent to 'harrow the valleys, or be bound with a band in the furrow'—when it 'laughs at the multitude of the city, and regards not the crying of the driver'—when, refusing absolutely to make ropes out of sea-sand any longer, it sets to work on statue-hewing, and you have a Pluto or a Jove, a Tisiphone or a Psyche, a Mermaid or a Madonna, as Fate or Inspiration direct. Be the work grim or glorious, dread or divine, you have little choice left but quiescent adoption. As for you—the nominal artist—your share in it has been to work passively under dictates you neither delivered nor could question—that would not be uttered at your prayer, nor suppressed nor changed at your caprice. If the result be attractive, the World will praise you, who little deserve praise; if it be repulsive, the same World will blame you, who almost as little deserve blame.[1]

Wuthering Heights, of course, stems from the Gothic novel, and Heathcliff could scarcely have existed had not Byron so persistently turned his profile to the world. The extravagance of its origins is strong in Emily Brontë's book, and she safeguards herself from ridicule by the indirectness of much of the narration—just as Hawthorne does in *The Scarlet Letter* by the elaborate account of his reconstruction of the story from a documentary record, and as Melville does in *Moby Dick* by a frequent lapsing into facetious comment. But the frenzy of Heathcliff over Catherine living or dead is as valid and as absurd as Tamburlaine's futile assertion of his power or Crashaw's girl child setting out on the path of martyrdom.

But this is indeed only one side of the picture. Not only some comedies but writings of many different kinds assert or imply a norm of rationality, of decorum, of a tempered brotherly love, of an unenthusiastic acceptance of the community's will and of traditional belief—a norm which men would be happier, it is implied, to live by. Happy

[1] *Wuthering Heights* (Everyman's Library), 1935, pp. xxvi–xxvii.

endings in comedy are not always ironic, and in tragedies the end commonly brings a resumption of the quieter current of life, with the suggestion of a temporary and impoverished security. In E. M. Forster's description of Beethoven's Fifth Symphony in terms of programme-music, the goblins have withdrawn at the symphony's close: they may always come back, but for a while we are free of them, free to imagine ourselves living in a Barsetshire where loss and grief are endurable and not too piercingly contemplated and where man can count up his undeniable small accomplishments. In this instance, however, and in all major literature, as well as in a considerable amount of minor literature, the assertion of a norm co-exists with an implication of the norm's fragility, of indeed the absurdity of postulating any pattern in a world we only darkly know. This coheres with a view already expressed—that the type-situation in literature implies the obverse of what is immediately propounded.[1]

It seems worth illustrating this in relation to the law-court scenes that are so frequent in Elizabethan and Jacobean plays. Now, the law represents one of the most important and most ambitious of human undertakings. It is also an essential part of every form of society. In *Of the Laws of Ecclesiastical Polity* Hooker could see the devising of laws as an attempt to codify the principles that God's will had at that time made manifest to men, either through scripture or through the promptings of conscience or through inheritance from their forefathers. Thus he could speak of the laws which men had made for themselves along with God's laws for Nature, and could offer praise for this total Law as the harmonising principle of God's world:

Wherefore that here we may briefly end: of Law there can be no less acknowledged, than that her seat is the bosom of God, her voice the harmony of the world: all things in heaven and earth do her homage, the very least as feeling her care, and the greatest as not exempted from her power: both Angels and

[1] See above, 'A School of Criticism', p. 40.

men and creatures of what condition soever, though each in
different sort and manner, yet all with uniform consent,
admiring her as the mother of their peace and joy.[1]

Yet when the men of Hooker's generation came to put on
the stage the administering of the law, they almost uniformly
showed courts as places of stupidity and vindictiveness and
frequently of corruption. Almost always the court is unable
to arrive at the truth through its own investigations: it has
to abdicate in favour of one of the parties in the case who,
by admission of his own guilt (as with Jonson's Volpone or
Tourneur's D'Amville in *The Atheist's Tragedy*) or by the
revealing of hitherto withheld evidence (as with Helena
in *All's Well that Ends Well* and the Friar-Duke in *Measure
for Measure*) or by the interpretation of the law in a way that
the presiding judge has not anticipated (as with Portia),
is enabled to control the course of events. Frequently the
figure of the judge is brought to ridicule, as with the King
of France in *All's Well* and the judges who find themselves
befogged at Volpone's trial. We can see in the Jonson
comedy a judge of Venice flattering and favouring the now-
rich Mosca whom yesterday he had despised, and thinking
to make him his son-in-law. In Webster's *The White Devil*
the prisoner Vittoria can protest that the Cardinal is
acting as her accuser as well as her judge and can swing our
sympathy to her because we know that, guilty as she is, this
court is not fit to give sentence. And when the revelations
have come in the ways indicated, the courts are ever ready
to take advantage of their power. We shall remember how
Volpone and Mosca are sentenced to a life of continuous
pain.

Less remembered is a sentence imposed on the bawd
Cataplasma and her male and female servants in *The
Atheist's Tragedy*. In her house a man has been killed (as
a man was alleged to be killed in Mistress Quickly's house,

---

[1] *The Works of that Learned and Judicious Divine, Mr. Richard Hooker,*
Oxford, 1850, I, 228.

with equally summary and perhaps merely convenient punishment). It was an affair of accident, as the court freely admits. Yet here is the sentence imposed on all three:

> Your goods, since they were gotten by that means
> Which brings diseases, shall be turn'd to th' use
> Of hospitals; you carted through the streets
> According to the common shame of strumpets,
> Your bodies whipp'd till with the loss of blood
> You faint under the hand of punishment.
> Then, that the necessary force of want
> May not provoke you to your former life,
> You shall be set to painful labour, whose
> Penurious gains shall only give you food
> To hold up nature, mortify your flesh,
> And make you fit for a repentant end.
>
> (V. ii. 30–41)[1]

The arbitrariness of this is underlined when immediately afterwards the hypocritical Puritan Languebeau Snuffe, whose guilt is far greater, is merely told to give up preaching and to return to his earlier trade of candle-making. And when the arch-villain D'Amville enters in distraction over the deaths of his sons, the Judges invite him to join them in the administering of the law upon the innocent Charlemont and Castabella. The truth is made eventually plain only through D'Amville's insisting on acting as the executioner, dashing out his own brains as he swings the axe, and then making (despite physiological unlikelihood) a full confession before he dies. The court would be helpless without this intervention by God's hand, as on other occasions it depended on help from human wit.

On one famous occasion, however, there is no revelation needed: the facts are known, and the man who passes sentence does so with some leniency if with sternness of word. The judge in this instance is the newly crowned Henry

---

[1] *The Atheist's Tragedy*, ed. Irving Ribner (The Revels Plays), London and Cambridge, Mass., 1964.

V banishing Falstaff. Shakespeare demonstrates the public
necessity of this. The King must not have as a companion
a man who declares that any man's horses may now be
stolen, for the laws of England are at his commandment.
Nevertheless, the picture we are given is not of Order
banishing Disorder but of a particular human being, a
skilful opportunist of a king, thinking to put down or
control the embodiment of universal anarchy. Disorder
cannot be effectively banished from the court or from
England or the world so long as the moon circles the
earth and human affairs are subject to mutability. The
essence of the scene—aside from its immediate and
convenient political effect—is Henry's arrogance and
complacency. He joins hands with the King in *All's Well*
and with the Duke in *Measure for Measure* who took over the
judge's seat when he had played out his part as accuser when
still wearing his disguise.

If men are to live together, some man must take on
himself the administering of the law, but to do so involves
a terrifying arrogance—tolerable only when that fact is
faced and the judge recognises his own unfitness. There
is no Elizabethan or Jacobean play where a judge shows
that measure of humility. Typical is the Senator in *Timon of
Athens* who discards the very principle of mercy—'Nothing
emboldens sin so much as mercy' (III. v. 3), he says—and
thus on principle refuses to heed a plea of extenuation.
And a man's self-confidence in the assumption and employment of power is grotesque and absurd whether he is seen
as a creature of God disposing of the lives of others or is
seen in a universe partially illuminated, in a single corner,
only by the candlelight of human observation. Shakespeare
and his contemporaries saw man in both these ways, and in
this particular regard the two views worked complement-
arily.

Great literature does on occasion make non-paradoxical
assertions concerning things that we are invited to recognise
as good. It is, we have seen in the previous section of this
book, the way of the Grand Style to function like that. But

we have also seen how, even within that mode, the affirmation may on consideration look frail. The situation is
more precarious in other writings: as noted earlier, Robert
Ornstein has drawn attention to the smallness of the
consolations for old age—'honour, love, obedience, troops
of friends'—that Macbeth has lost through his murder of
Duncan;[1] and certainly the casual placing of 'love' between
'honour' and 'obedience', the indiscriminateness implied
in the word 'troops', do not make the autumnal picture
one to rejoice the heart. In *King Lear* loyalty and affection
and a wide sympathy and understanding are seen as
sporadically existent and approvable, but they are subject
to time and human malice: they endure only for a moment
and they do not impose themselves on society. And, in
writing of a very different sort, we can observe Pope in
1715 inserting the speech of Clarissa in the fifth Canto of
The Rape of the Lock. This was, he said in a note, 'to open
more clearly the MORAL of the Poem', and he described it
as parodying a speech in *The Iliad*. As parody, it disavows
seriousness, and the lightness of its touch is appropriate
to the poem in which it is inserted. Yet it is insistent on the
pathos of mutability:

> Oh! if to dance all Night, and dress all Day,
> Charm'd the Small-pox, or chas'd old Age away;
> Who would not scorn what Huswife's Cares produce,
> Or who would learn one earthly Thing of Use?[2]

And as a means of preserving one's poise while one con-
fronts the ravages of the years, Clarissa urges upon Belinda
and the rest the cultivation of 'good Sense' and the preservation of 'good Humour'. The words are wise enough, but the
consolation is frail and the counsel rarely accepted. We are
not surprised when Pope tells us that the speech won no

---

[1] See above, '"The Servants Will Do that For us"', p. 11: cf. also
'The Dark Side of *Macbeth*', *The Literary Half-yearly* [Mysore], VIII
(January and July, 1967), 27–34.

[2] Canto V, ll. 19–22; Pope, *ed. cit.,* p. 238.

applause, that Belinda frowned and her friend Thalestris declared Clarissa a prude. Of course, in literature of a more manifest ambition and a firmer faith, the affirmation may be stronger, the promise larger. We may be told that nothing is here for tears when God's champion triumphs over God's enemies, that in his will, as Dante has it, is our peace. But even in such cases the affirmations rest on the great paradoxes of theism: the tears—and indeed, for Samson, the blame—are only with difficulty to be kept back, the peace of submission is expensively won.

Thus the picture that literature generally offers of man's affairs and of the cosmos is phantasmagorical, unreasonable; and this is true whether it is concerned with the aberrant individual (the Tamburlaine, the Teresa, the Heathcliff) or with society's normal conduct of itself; the consolations offered are either frail and subject to time or formidable in the challenge they present. But not merely does literature envisage a world that in its available condition is absurd. The absurdity extends also to the origins and the mode of composition. The impulse to start writing comes ultimately, as we have repeatedly seen, from the need to escape from the flux of experience, to isolate a certain part of the flux and thus to perceive it as an entity, a 'thing'. This we all do, whether writers or not, employing what Coleridge called the Primary Imagination.[1] But the isolation will be incomplete unless we have a means of objectifying the thing isolated, of allowing us to place the thing outside the stream of our consciousness: perception by itself must be momentary, subject always to the modifying pressures of other perceptions and of general notions. And so, if we can, we use words or whatever other means are available for the task of objectifying. But the words bring along with them associations with other acts of objectification which have already been made through them.[2]

The new entity can be isolated only through the use of

[1] *Biographia Literaria* (Everyman's Library), reprinted 1910, p. 159.
[2] See above, 'Comedy in the Grand Style', p. 54.

a method which itself prevents isolation: in the act of being rescued from the flux, the 'thing' takes its place in a new flux, the stream of literary tradition. This is made plainer if we consider the history of literary Kinds. It is true that from time to time new Kinds are invented, but usually as a modification of already existent Kinds. In the drama of the Renaissance the pastoral play and the history were new forms, but with associations respectively with the older pastoral poem and the long-established tragedy and comedy. When the novel became a recognisable Kind or group of Kinds, it could in retrospect be seen as stemming from epic or romance or, later on, the dramatic forms. What this suggests, however, is that at any given time a writer will almost certainly, for a given act of objectification, cast his writing into an available mould.[1] Yet the essence of his activity depends on the uniqueness he is conscious of in the 'thing' he is perceiving.

Admittedly it is a uniqueness that goes along with a degree of kinship with other 'things'. But almost at the outset of the writer's task the idea of uniqueness is disregarded, as the poet chooses, for example, the sonnet-form, or the dramatist decides to write a comedy. And he customarily needs to do this in order to canalise, to escape from the flux of experience. Sometimes, indeed, he will go further, and write a deliberate variation on an existing work or type of work, as Joyce used the *Odyssey* structure as a scaffolding for *Ulysses*, as Pope fitted his mock-epics into the epical pattern, as Byron used those same models from antiquity as a point of immediate departure for *Don Juan*, as modern dramatists—Anouilh, Sartre, O'Neill, André Obey—have made their variations on particular Greek plays. Sometimes the variant is close to the original, sometimes deliberately remote, but in every instance the isolation of the thing perceived, the recognition of its uniqueness, is at once compromised by the association which is made.

[1] See above, 'Comedy in the Grand Style', pp 54–5.

84

It is true that some Kinds are more comprehensive than others, that, for example, the special type of tragedy that Shakespeare wrote and that Marlowe anticipated—a mode of writing eschewing the Grand Style and all its manifest strength—simultaneously presents human greatness and littleness, and interrelates affirmation and dubiety. Nevertheless, this is a kind of writing as precarious as any and is indeed rarely attempted with success.[1] Shakespeare has given us only a handful of plays that we can regard as fully belonging to the tragic Kind: in his early career his *Titus* is marred by a rhetoric which stultifies the vision of evil, his *Romeo and Juliet* ultimately expends itself in moralising; and in the years of his maturity it was easy to lose faith in the minimal, qualified affirmations necessary to tragedy, as he seems to do in *Timon of Athens*. It is not surprising that the main achievements of modern drama have been in what O'Neill called 'a big kind of comedy that doesn't stay funny very long',[2] the kind of comedy that we see in *Peer Gynt* and *The Wild Duck*, in *The Cherry Orchard* and *The Three Sisters*, in *Juno and the Paycock*, and in O'Neill's own *The Iceman Cometh*. Christopher Fry once said that in the end our intuition guides us to comedy, implying that tragedy makes premature assumptions about the nature of the world, imposes a too rigid pattern upon the 'thing' embodied.[3] If there is a justification for that remark, it must relate only to that kind of comedy which subsumes much of the tragic, not to the kind that depends on a disregard, or a softening, of the terrible. But, however inclusive a Kind may be, the writer's almost inevitable dependence upon it negates his purpose. There is no isolation, no full preservation of a sense of uniqueness, but merely a transposition from one form of flux to another.

---

[1] Its peculiar quality is to some extent explored below, in '*Catharsis in English Renaissance Drama*', pp. 145–6.

[2] Crosswell Bowen, *The Curse of the Misbegotten: A Tale of the House of O'Neill*, London, 1960, p. 311.

[3] 'Comedy', *The Adelphi*, November 1950, pp. 27–9.

This essay began by drawing attention to particular pieces of writing which, with the full intention of the writers, invoked a recognition of absurdity. But I have tried to argue that these are simply the moments when, as in Chapter LIX of *Moby Dick*, the Great Squid floats briefly to the surface and its shocking amorphousness is revealed. More generally this is less apprehensible, but the basic character of the object—or of the activity, for what object is not, for the perceiving mind, an activity?—is always the same. All serious writing may be seen as absurd, both in the picture of human experience it overtly or surreptitiously offers and in the mode of its composition. Great writing may come into existence when the writer knows that these things are so.

Chapter 5

THE SHAPING OF TIME

U NTIL this century, the action of a drama was plotted in
what we can call a straight line. If we saw Event B on
the stage after Event A, we were expected to assume that it
came later in time. Aristotle's demand, of course, was that B
should arise out of A 'by probability or necessity',[1] but even
the most carefree and romantic of playwrights would adhere
to the principle that succession in presentation implied
succession in action. What we get from such writing is a
series of present moments correspondent indeed to our normal and superficial concept of actual experience. Memory
and anticipation are not ruled out, and memory plays a
particularly important part in those dramas where the Unity
of Time is observed: Tamburlaine and Henry V can begin by
dreaming of the conquests they will later achieve; *Oedipus
Rex*, *The Wild Duck* and *The Tempest* can use memory so fully
that, while the explicit action moves always onward, the
spectator's mind is busily constructing a past which is no
longer moving because complete. In these last instances it
is worth noting that the past in such a play is commonly
removed from the play's action by a substantial tract of years.
The past that matters for Oedipus is comprised of events that
terminated with his marrying Jocasta and becoming King;
for Gregers Werle and Prospero the special burdens they
carry are not related to what happened only yesterday. Thus
commonly—though not uniformly, the linked history plays
of the Elizabethan years being among the exceptions—there
is a sharp sense of difference between what is presented
and what is referred to as anterior; and almost always the

[1] *The Poetics*, Chapter IX.

principle of temporal succession is adhered to. It is true that it was possible to use a framing device, with for example the dramatised action functioning as something recalled by a narrator—by Gower in *Pericles*, by Madge in Peele's *The Old Wives' Tale*, by Bohan in Greene's *James IV*. Yet I know of only one instance, even in such cases, where time is treated lightly inside the action thus framed. In *James IV* Bohan is presenting the play to Oberon in order to demonstrate the condition of things in the world that long ago made him turn recluse: the dramatist at one point allows Oberon to intervene in the action to save one of Bohan's sons from execution. Here magic is enabled to transform the past, but Greene's example was not followed in the plays of his successors, and thus remains a curious exception to an otherwise uniform adherence to linear time in dramatic action.

This may at first sight appear strange. The Elizabethan dramatists are notorious for playing certain tricks with time: they could move from midnight to dawn within a few minutes, as in the first scene of *Hamlet*, and they could use inconsistent references to duration, as most liberally in *Othello*. But there was dominant in play-writing a dichotomy between a completed past and a present that was shaping itself in relation to, and perhaps under the full control of, that past. This is in conformity with a total view that sees things ultimately 'working out'; the present is subsumed into the past; the idea of the past held by us in the present possesses, as we have seen, a character of completeness, but our actions in the present are continuously adding to the structure; a moment comes which crowns the work, when there is no more future, when the past is subsumed into a static present. This is symbolised in the terminal point of a play (or of any piece of narrative fiction cast in the same mould of thought); Time triumphs, as it is asserted in the sub-title of Greene's *Pandosto*; the whirligig of time brings in its revenges, as Feste sharply tells Malvolio at the end of *Twelfth Night*;[1] when

---

[1] How very sharply I have tried to indicate in *Twelfth Night and Shakespearian Comedy*, Toronto, 1965, p. 45.

Edmund in *King Lear* recognises that for him Fortune's wheel has come full circle, he employs the traditional emblem to signify that a pattern in his life, previously hidden from him, has now reached total being and become plain. But such terminal points in plays and early novels may be symbols of a larger idea.[1] For the individual man, the moment of completion is death, which gives a wholeness to his life, makes his total being ready for judgment—as Sartre has implied in *La Mort dans l'Âme*, the final section of his trilogy *Les Chemins de la Liberté*. And for the universe, as seen by writers for whom the Christian tradition is strong, there will also come a moment when succession evaporates, when becoming yields to being. Marlowe makes Mephostophilis anticipate such a condition:

> when all the world dissolves
> And every creature shall be purify'd,
> All places shall be hell that is not heaven.
>
> (v. 125–7)[2]

Thus we can understand that in a fictional presentation the principle of temporal succession would be normally adhered to without conscious questioning: Greene could make Oberon intervene in an event belonging to the past, but *James IV* is not a very serious play and the King of the Fairies could for once be permitted to enjoy a bizarre extension of the powers normally credited to him. Such a thing could not happen in a purposeful holding up of the mirror to nature, and even in comedy of the lightest sort it remains, as far as I know, a unique departure.[3]

[1] In 'Shakespeare and the Idea of the Future', *University of Toronto Quarterly*, XXXV (1966), 213–28, I have suggested that the use of such symbols of finality in drama does in practice go along quite often with a continuing preoccupation with the unfolding of event.

[2] *Doctor Faustus*, ed. John D. Jump (The Revels Plays), London, 1962.

[3] There is a hint of something similar in Medwall's *Fulgens and Lucres*, where the players A and B decide to take part in the action, which belongs to a period before their time, but the action is not substantially modified by their presence.

89

But suppose our view of the world is of a different sort, not one in which becoming yields ultimately to being, in which a pattern is finally established, but one in which a man's death is no more and no less than a single incident in his life, in which time does not triumph or bring in its just revenges—in which indeed hero and villain and common man are equally 'caught' in what we call the 'end', in which justice does not exist except in the mind of man, in which causation itself is a concept to be handled gingerly. We may hear of a friend dying in a painful and grotesque fashion, and our immediate reaction is to let that fact colour our whole sense of his being: for the moment each step in his experience seems only to have been bringing him nearer to that final happening which is felt to shape all that went before. But when the initial shock of the news has passed, we may work closer to a freedom from that traditional response, and our recollection of the man as he was in separable moments before death will assume something approaching autarchy. It is even, perhaps, possible for a man about to die to see his death thus, as only one incident among others. Katow in Malraux's *La Condition humaine*, walking towards the railway-engine where he was to be thrown into the furnace, could say to himself: 'Allons! supposons que je sois mort dans un incendie.' That of course exhibits a degree of self-command that makes us catch our breath, but the point that is especially important here is that, not only was there an absence of justice in Katow's end (and, in the writer's view, nothing beyond that end in the way of compensating experience), but the end itself did not negate Katow's life. He died terribly, but that is a separable fact from his having lived, and with a kind of splendour. It is indeed our normal reaction to the death of a tragic hero, but here we are taken to a limiting point because of the insistence on physical pain. Without that insistence, such a view of things can lead to a complacency as offensive as any facile notion of ultimate compensation. In Jean Anouilh's play *L'Alouette* the story proceeds to the burning of Joan of Arc, and then the commentator asserts that Joan's story

was nevertheless a triumphant one and the play will therefore end with a re-enactment of her moment of success when
the Dauphin was crowned at Rheims: so time in the theatre,
as often nowadays, is made to turn backwards and we see one
of the play's earlier scenes repeated, we see Joan smiling
once more, and we leave the theatre glad for the success of
French arms in rebutting the invader. On the way home we
may resent the trick has been played upon us. The situation is indeed different with Malraux: he does not spare
us a sense of the agony awaiting Katow, and in another of
his novels, *La Voie royale*, he makes one of his characters,
watching the painful death of his friend, feel a longing
that there were indeed gods to whom he could cry in his
anger that no hope of heaven, no promise of reward, can
justify the end of any human life. For Malraux, death is
terrible because it brings to an end and because the struggle
against non-being can be agony; but it does not govern
the pattern, it does not invalidate the life that has been.
Here the finely articulate Malraux gives utterance to a
view of things which in our century has gained ground and
which has incidentally led some dramatists to a freer
use of temporal patterns than their predecessors thought
of.

The novel in its earliest phases was almost as fully wedded
as the drama to the principle of succession. Philosophically
there could be no difference between the two modes: in narrative fiction time had its triumphs as regularly as in tragedy
and comedy. But for technical reasons there is an important
difference. In the drama a clear distinction exists between
what is narrated and what is presented: the former is conveyed through a set speech (either by a chorus-figure or by
one of the *dramatis personae*) or on occasion through an inter-
change between two or more speakers, while the latter is
acted out before our eyes; the recollected and the immediate
belong in different worlds of discourse correspondent to the
separate notions of past and present. But in the novel all is
narrative: there commonly is an immediate story-line, but
when one of the characters talks about the past or where

the omniscient narrator (like a chorus in a play) interrupts the story-line in order to apprise us of things we need to know concerning events that have led up to those with which we are to be primarily concerned, the interposed narrative will not necessarily differ in its mode of presentation from the main story-line itself. Thus the story of the Paphlagonian King in Book II, Chapter X, of the *Arcadia* and the story of Mr. Wilson in Book III, Chapter III, of *Joseph Andrews* present events of the past in a mode only partially distinct from that of the adventures of Musidorus and Pyrocles or of Parson Adams and Joseph. There are indeed degrees of distinction in such cases, for in Sidney the narrative is sufficiently short for us to remain conscious of the presence of Musidorus and Pyrocles, who are listening to it, and Fielding goes so far as to make Adams interrupt Wilson from time to time or at least silently register his response to what is being narrated. To that extent a past-present dichotomy is maintained in a fashion inherent in the dramatic writings contemporaneous with these novels. But when the inset narrative is substantial in length and the novelist allows it to be self-contained, the distinction evaporates. Moreover, a writer of narrative will frequently have more than one character as a centre of attention, as Spenser has when Una and her knight are separated, as Cervantes has when Sancho goes to his island: he will then commonly find himself in the position of having to travel back in time in order that we may learn what has happened in the interim to the character who for a time has been neglected. 'We last saw our hero when he was taking a sad farewell of the beauteous Isabel, little knowing of the danger that was to threaten her at home while he was far away.' We have already been given a narrative of Isabel's misadventures, and now we go along with the hero until once again we get to the point of time, or thereabouts, where she in her turn has been left. This of course can be tedious, and many a reader of lesser Victorian fiction has yawned at meeting such words as have here been put together to illustrate the use of the device at its most elementary. Even in this small way, however, the novel's long-established freedom in its use of time is evident.[1]

There is, moreover, a subtler distinction between drama and narrative fiction which is relevant here, and that concerns the nature of a 'present moment' in the two modes. In drama what is present is and must be something discrete. It is a man stabbing or kissing or dying, a man saying the words 'To be or not to be'. In the novel there is inherently no such sharpness of effect. For one thing, all is in the past. The indications of action are nearly always in the past tense, unlike the stage-directions which a reader of a play is given. For another, the presence of the narrator, even at his most reticent, is brought to mind at least in such necessary formulae as 'said John'. And there is the use of the character-ising adverb, or the frequent indication that what a character is saying is in complex relation with his condition of mind. At every instant there is a doubleness of viewpoint: the experience presented is seen both through the eyes of the character or characters involved (and for him or them it is immediate, present) and through the eyes of the narrator (for whom it has taken its place in a total fabric of narrative and thus belongs to the past). Clearly the drama approaches the condition characteristic of the novel when a framing action is used, but even then the fact that we watch real people acting makes what they do and say possess a high degree of self-containment. In the novel we are always, to a greater or lesser degree, aware of the present as simul-taneously past. We are aware too of the complexity of each moment, of its associations within a series. Drama at its simplest has, as emphasised already,[2] the quality of a rite, in which each act and word has a hieratic function. Narrative,

---

[1] In a very few cases the drama at an early stage used a novelistic technique of this kind: Thomas Heywood's *The Four Prentices of London* (*c.* 1600) provides an example; double or multiple plots in Elizabethan and Jacobean plays are usually so separate that we are hardly conscious of the need to 'catch up'. The basic dramatic mode before the present century inherently remains a straightforwardly progressive one.

[2] See above, ' "The Servants Will Do That For Us" ', pp. 21–2.

on the other hand, puts us into the flux. Among writers earlier than the late nineteenth century, it is Sterne who avails himself most fully and self-consciously of the freedom that narrative thus allows. He can expand the moment, prolonging it by holding its many facets before us in turn; he can reverse his tracks, making the past once again present; he can insist on the complexity of response that a group of characters may display to a single happening; he will delight to break his narrative to add his own particularising and generalising comments.

But in the writing of recent generations even Sterne's degree of freedom has become an established thing. So much so, indeed, that the new mode can be itself mocked in a novel, as by Günter Grass's narrator in *The Tin Drum*:

> You can begin a story in the middle and create confusion by striking out boldly, backward and forward. You can be modern, put aside all mention of time and distance and, when the whole thing is done, proclaim, or let someone else proclaim, that you have finally, at the last moment, solved the space-time problem.[1]

Grass's fun is legitimate but does not alter the fact that the novel has shown itself to be the mode most appropriate to the modern consciousness of the nature of living. Drama has won a large measure of freedom, but its inversions of the temporal sequence have often about them something of the perversity of an anti-ritual. The novel and the film, on the other hand, have become the basic media of the twentieth century when it is taking itself most seriously in the field of fully articulate expression. The drama is capable of strong and poignant gestures in the deliberate rejection of its historic mode: indeed it remains fully necessary in its new function of explicit denial. Nevertheless, novel and film are predisposed to temporal freedom, and I wish now to discuss two major examples of a free manipulation of event— Conrad's *Nostromo*, first published in 1904, and Malcolm Lowry's *Under the Volcano*, first published in 1947.

[1] *The Tin Drum*, tr. Ralph Manheim, Harmondsworth, 1965, p. 13.

In common they have a setting in America south of the United States—*Nostromo* in an imaginary republic called Costaguana, facing the Pacific, *Under the Volcano* in Mexico—and a concern with the tragic event. In other respects it might appear they are as far apart as *The Winter's Tale* and *The Tempest*—the one ranging freely over a wide tract of time, from that moment when the Gould family were first made to assume the burden of the silver mine to a future only hinted at, when the Goulds' success is once more brought into question; the other giving us (after a retrospective opening chapter) the anguished events of a single day. Of course, the Shakespeare analogues are not exact: *The Winter's Tale* follows the line of temporal progression, *Nostromo* repeatedly comes back to the same moment of crisis before finally taking off into more or less plain narrative; the conclusion of *The Tempest* is poised between the completed and the open-ended effects, displaying and simultaneously questioning the point at which time has triumphed,[1] while the opening chapter of *Under the Volcano* takes away from the Consul's death, at the end of the novel, some part of its finality. Nevertheless, just as the two Shakespeare plays combine strong technical differences with a similarity in mood and viewpoint, so the two novels exemplify different ways in which the twentieth-century writer may show his readiness to treat time with disrespect, to see death as an incident, to assert the validity of the thing that is as, and because, it happens.

The first chapter of *Nostromo*, like the beginning of *The Return of the Native*, gives a static description of the setting: the Placid Gulf, with the 'insignificant' Punta Mala on one side and the peninsula of Azuera on the other (one ominous in name, the other through its legend of men lost in a search for treasure), the islands called the Isabels forming a barrier against the open sea, and the location of the Sulaco plain and its relation to the republic of Costaguana. Isolation is emphasised, the mercantile

---

[1] The matter is explored a little in 'The Structure of the Last Plays', *Shakespeare Survey* 11 (1958), pp. 19–30.

adventurers being kept away by the prevailing calm of the gulf and the mountains beyond the plain preserving a quasi-independence for Sulaco; but the hints of ill omen on either side of the gulf, the gauntness of the islands, and the implicit suggestion that steamships, soon to replace sail, will not be so easily deterred—all mark this as a setting for tragic events. It seems we are to witness the destruction of a Utopia. But we soon learn that that is to see things too simply: Costaguana is a republic already torn, subject to the historical tensions of the nineteenth century even before the Placid Gulf ceases to be a deterrent; dictatorship, federalism, liberalism, take their turns in the country's mode of government; semi-isolation is no guarantee against violence, corruption, intrigue; all it effects is to keep these things local, domestic.

In the second chapter we are taken to a point in Costaguanan history to which the novel returns again and again—the defeat and flight of the liberal head of state Ribiera. The novel is divided into three Parts—'The Silver of the Mine' (giving the ideas of both romance and 'material interests'), 'The Isabels' (forbidding in view of the opening account of the islands, dramatically the turning-point for Nostromo and Decoud, suggestive in their name of the part that women play in the destruction of both men), and 'The Lighthouse' (ironically symbolic of the new link with the outside world, and the scene of Nostromo's end—the end of the complete individualist being appropriately part of the narrative which brings Sulaco into the North American orbit). When Sulaco becomes independent of the rest of Costaguana, it is received into the larger mercantile world, becoming a valuable scrap in a larger series of dependencies. At the end of the book we are reminded that this too is not a static condition, that inimical political forces are beginning to gather strength, that the San Francisco millionaire will not always be the mine's remote landlord.

Conrad's characteristic movements backwards and forwards in time are profoundly functional here. The story could have been told straight on, the rise and fall of rival powers being exhibited in normal progression, the people growing old and in turn dying, the romantic, infertile marriage of the Goulds slowly withering, Antonia suddenly diminished by Decoud's death, Decoud the visionary liberal being disposed of some time before the tougher, ideology-free Nostromo found that his small (though large for him) raid on 'material interests' did not free him from time and chance. The book is dedicated to Galsworthy, and in a sense the people of Sulaco are Conrad's analogues to the Forsytes that Galsworthy was later to chronicle, struggling with their load of silver, living and dying within the context of an historical process. But an analogue closer in quality to Conrad's work is Thomas Mann's *Buddenbrooks* of 1902, which, like *The Forsyte Saga*, uses the method of straight narrative to show the operations of time. A major book, even in the twentieth century, does not have to juggle with the time-scheme.

Yet we have seen that the point in Conrad's narrative to which we are first taken is repeatedly returned to. After Part I, in Chapters II–IV, has told us of Ribiera's defeat and the good deeds accomplished by Nostromo at that time, the remainder of Part I moves backwards. First, in Chapter V, we learn of events eighteen months before, when Ribiera's government was precariously shedding the light of liberalism over the country and the Goulds' mine, and then in the remaining three chapters of Part I we are taken further back, to the brutal dictatorship of Guzman Benito, the saddling of Gould's father with the burden of the mine, the resolution of Charles Gould (romantically taken at the time of his courtship) to make the mine flourish, his time of growing success with North American help, and then in Chapter VIII, at the end of Part I, we are back to the story of Ribiera's defeat. Part II has also eight chapters, and the first seven of these likewise move towards the same moment of crisis in Costaguanan fortunes as had been the subject of the opening and concluding chapters of the narrative in Part I. Only Part III moves definitively forward beyond that point, though often turning back upon itself to juxtapose present with past more

97

ironically, and, with particular force, giving us Decoud's death (though it happened earlier) as part of the narrative leading to the fulfilment of his dream of independence for Sulaco, an independence achieved only through an entry into a larger subjection. The retrospects of Parts I and II are different in tone and theme. That of Part II is more intimate, concerned far more with personal relations: the small, gracious figure of Mrs. Gould in Part I becomes more fragile, more poignant and resigned (yet stronger too) as her husband's concern with the mine grows under our view from a romantic enterprise (appropriately linked with his wooing of her) to an obsession that she cannot share, that puts a silence between them. And we come closer to Antonia and Decoud and Dr. Monygham and, of course, Nostromo in Part II, sharing with them all that strange half-calm in Sulaco that was made possible while Ribiera ruled and the implications of North American patronage were not fully evident. The relation of the book's three parts provides the basis for a superb structure, the lingering with the past in the first two-thirds of the narrative hinting at a reluctance to plunge into the characteristic world of the twentieth century, the free movement between present and past placing an insistence also on the *fait accompli*, on the entrammelling of fully alive individuals within a partially hidden progression which, however enterprising they may be (Charles Gould in legal, Nostromo suddenly in illegal, adventure, Decoud in his political vision and his love), is not within their control. The progression forward in Part III coheres with our experience that a movement after a crisis commonly seems a thing long delayed (as in the 1930s the effective beginning of the Second World War, or in our personal lives the change that is long anticipated but seems indefinitely postponed), and then suddenly the barriers are down and events surge on. The book's political and mercantile happenings are part of the world's pattern at that time, and, in echo of Tolstoy, Conrad makes his smaller Napoleons into marionettes controlled by something that operates of its own momentum, whether because it is so destined or because the cards have fallen at that moment in that way. Yet these marionettes are as fully alive as anyone we know.

It may still be asked why it is Ribiera's defeat that is the point so often returned to. That, surely, was the end of a dream, the dream of liberal self-government for a community, with the fantasy informing it that violent men can be reasoned with and thus controlled, that financial help from outside can be wholly beneficent. The aftermath of the defeat is first a period of unleashed violence (within which Nostromo's crime is understandable and congruent) and then a period of ominous success for the Gould enterprise and the power that from afar shapes it. The irony of Part III is made clear in Chapter XI, where Mrs. Gould comes to realise that

> There was something inherent in the necessities of successful action which carried with it the moral degradation of the idea.

To this is juxtaposed, in the preceding paragraph, her own vision of the good life:

> It had come into her mind that for life to be large and full, it must contain the care of the past and of the future in every passing moment of the present. Our daily work must be done to the glory of the dead, and for the good of those who come after.

That is a possibility for Mrs. Gould; hardly for Antonia, whose life after Decoud's death is an empty looking back and a devotion to good works in the present; not at all for Charles Gould, with his eye fixed on building the fortune of his mine; or for Dr. Monygham, for whom recollection is torment, the future of no account, and the present a matter of painful, wholly disinterested love for Mrs. Gould. She, as much as Nostromo, is in the book's centre, the figure of affirmation who, with a barrier between her and the husband she loves, quietly defies process in her vision of a present which is lived in full respect for what precedes and what follows. Whatever the past has been and the future will be, they are for her things which the present, otherwise barely existent, must serve. But the goodness is in the service: it

PROPERTY OF
UNIV. OF ALASKA LIBRARY

will not bring good into new being or banish grief; the dead must be honoured despite the anguish of memory; the world and its people must be cultivated, though nothing tarnishes more obviously than the realisation of an aim.

This is no occasion for an adequate study of this major novel, but it is hardly possible to turn from it before noting some of the recurrent motifs that help to give depth to its treatment of human beings within a tract of time. The fact that old Viola had served with Garibaldi links the story with that historical example of a triumph tarnished. The use of the nickname 'Nostromo' by the upper-class liberals is a hint of possessiveness that goes along with Charles Gould's exploitation of the mine, his mere acquiescence in his wife's love, his unaware betrayal of Sulaco into the power of the old man of San Francisco. But of course the nickname also signifies that in the larger sense, like Lord Jim, Nostromo is 'one of us': he belongs with us all, not merely to those who patronise him. The pathetic cooperation of the outlaws in 'freeing' Sulaco counterpoises the dependence of the Goulds and their friends on 'our man' the Capataz. The presence of Hirsch in the boat carrying its load of silver, Decoud's and Nostromo's astonishing avoidance of capture, the chance that Nostromo was misunderstood by Viola when he went to ask for Giselle as his wife, the chance that Viola shot Nostromo in mistake for Ramirez—these things, like so much in the book, defy 'probability or necessity', though they are all congruent with the people concerned. Even the moment of panic which made Nostromo speak ambiguously to Viola when he said 'I have come to ask you for . . . my wife!' fits within the context that has been established: Nostromo the thief is subdued in the presence of the single-minded fighter for republican liberty. In the concatenation of events it is chance that appears to operate, while forcing on us the impression that all is as it must be, thus giving to the whole concept of time a more terrifying quality.

Lowry's *Under the Volcano* is at first sight a book with a much narrower compass. Apart from Chapter I, the events directly presented occur between an early morning and the same day's dusk. There are only three characters whose fortunes we follow throughout: the Consul (no longer a consul), his wife Yvonne, his younger brother Hugh. They move about in or near the town of Quauhnahuac, never out of sight of the twin volcanoes that dwarf the men at their feet and contain within themselves the threat and promise of destruction—twins, male and female, brother and brother, gods yet men too in their joint rooting in the earth, their joint reaching away from it, their shared element of suppressed fire. In the main line of narrative there is no movement back and forth, apart from the retrospective Chapter I. The day's action occurs in 1938, but in Chapter I it is a year later than that: Dr. Arturo Díaz Vigil and M. Jacques Laruelle remember, as they drink together, the dead Consul, the dead Yvonne, the bereft Hugh. Yet Lowry is strangely inexplicit in this opening chapter. We are never firmly told about the deaths of the husband and wife until we come to the end of the book, and many a reader has gone on to Chapter II, and far beyond, in the belief that the main narrative follows, not precedes, the conversation of the two men with which the book begins and the musings of Laruelle that form the remainder of Chapter I. What Lowry may intend here is to give a fuller sense of life to his main characters by not simply asserting in advance what happened to them on their day of disaster, while at the same time making Vigil's and Laruelle's comments ominous enough and gradually letting them appear as retrospective. Moreover, the inexplicitness make the Consul and the others seem alive when they are being referred to in Chapter I, and this is a mode of defying the operations, even the fatal operations, of time. To have lived and to be the subject of anguish in recollection is in some sense to be living still. Perhaps, however, a very perceptive reader may get the time-scheme right at once: my own experience was not that, and is borne out by that of other readers I know.

In the rest of the book there are passages of recollection, of course, as there are in any substantial novel, but attention is centred on the events of the day. Outside this centre there exists an extensive periphery. Because it is 1938, the war in Spain is going on and the Nazi threat, strong over Europe, finds sympathetic reverberations among the small bullies of Mexican officialdom. What Lowry also does is to make us aware of the cultural complexity of a human being's store of memory. A man of our time may think in terms of Dante, for example, of the Elizabethan dramatists, of (especially in 1938) German films of the 1920s, early 'westerns' from Hollywood: strangely, the trivial and the masterwork can function similarly in providing the skeleton for thought, the interpretative symbol. When Yvonne and Hugh go riding in the morning, the great ravine they cross and skirt is called the 'Malebolge', because Dante provides the suggestions of depth and corruption and miscellaneous horror: it is this ravine into which the Consul is thrown, where he dies as he descends. Repeatedly we hear of the film *Hands of Orlac* being shown at the cinema in Quauhnahuac, a poor film with Peter Lorre acting in vain, a re-make of a German silent film in which Conrad Veidt had played: that, too, was poor enough. The Lorre version is already old, poignant therefore as a reminder of lost time, but it provides also a chance link with the 1920s, a time when the characters of the book were younger and Europe was not yet wholly dark: it acts therefore as a double filter for memory. Moreover, the melodramatic idea of the film—of the musician who lost his hands in an accident and had a dead murderer's hands grafted on in their place—becomes a symbol of the duality of the human being, making yet fearful of destroying, led on to destruction by the very instruments with which he makes. It is Hugh's love for Yvonne that leads him to betray his brother's trust; Yvonne's strong wish for a stable relationship with her husband that makes her despairing and promiscuous; the Consul's power of perception that makes him destroy his marriage, refuse Yvonne on her return, and finally acquiesce in the violence offered him. Any 'present' moment, when experienced in full consciousness, is a focal point not only for a patterning of memories but for an interpretation of both personal and political life in terms of cultural experience. So much so that one can never respond simply to what the moment itself offers. In Chapter IV Hugh and Yvonne are riding: the day is not yet hot, the country is gracious, the man and woman are in sympathy with each other:

> Earlier it had promised to be too hot: but just enough sun warmed them, a soft breeze caressed their faces, the country-side on either hand smiled upon them with deceptive innocence, a drowsy hum rose up from the morning, the mares nodded, there were the foals, here was the dog, and it is all a bloody lie, he thought.

Hugh goes on to recognise an appropriateness in the fact that this is a day of solemnity for the Mexicans, a day given over to honouring the dead, who 'come to life' for the occasion—like the memories of betrayal in one's own life. Even so, this is life at its best:

> Another thought struck Hugh. And yet I do not expect, ever in my life, to be happier than I am now. No peace I shall ever find but will be poisoned as these moments are poisoned.

A little later he almost forgets:

> Christ, how marvellous this was, or rather Christ, how he wanted to be deceived about it, as must have Judas, he thought—and here it was again, damn it—if ever Judas had a horse, or borrowed, stole one more likely, after that Madrugada of all Madrugadas, regretting then that he had given the thirty pieces of silver back—what is that to us, see thou to that, the *bastardos* had said—when now he probably wanted a drink, thirty drinks (like Geoff undoubtedly would this morning), and perhaps even so he had managed a few on credit, smelling the good smells of leather and sweat, listening to the pleasant clopping of the horses' hooves and thinking, how joyous all this could be, riding on like this under the dazzling sky of Jerusalem—and forgetting for an instant, so that it really *was* joyous.

He too could forget 'for an instant, so that it really *was* joyous'. The moment just after this, when he and Yvonne come to the brewery and drink the cold dark German beer and Yvonne hankers after buying the armadillo as a pet, is the book's idyllic moment, but *we* are never allowed to forget the recurrent personal betrayal, the burden of the mind's obligation to interpret, the large-scale disaster of the Spanish war, the coming extension of international strife, the hostility of nature to man. Only at a glance was the armadillo gentle:

> Each time the armadillo ran off, as if on tiny wheels, the little girl would catch it by its long whip of a tail and turn it over. How astonishingly soft and helpless it appeared then! Now she righted the creature and set it going once more, some engine of destruction perhaps that after millions of years had come to this.

Apart from forgetting, there is the sheer mystery of the past. Yvonne's is shadowy, linked with her success as a child-star in ephemeral films; Hugh's is barely given to us, but we know of the political urge that is on him, of his former relation with Yvonne, of the rift this makes between the brothers; the Consul is neither affirmed nor denied to have been guilty of a special atrocity in the First World War; the details of the first separation of Yvonne from her husband are never explicit for us. The past always has its obscurity and its defiance of credibility: he, or she, could surely not have done that, is commonly our response to what is alleged, and is as unbelievable to the one who has actually done it (whether heroism or ignominy is involved) as it is to those who know him on terms of intimacy. The roots of the present are things we feel we must refuse to accept or must leave as a question-mark.

The human mind has a further mode of self-tormenting. It is not easy on the best of occasions to forget that—'even now, now, very now'—things are happening in the world at large that mock at a present joy. So Hugh is recurrently conscious that, as he talks and journeys through the day, the battle of the Ebro is being fought and lost. But the special plague is not merely that we are doing nothing, can do nothing, to prevent the disaster far away: it is also that we believe each act of ours contributes to a total pattern that ensures the future: because one raises or puts down a glass, the battle of the Ebro will be lost; if one had acted, even in a trifle, differently, a different future would have been ordained. This is not rational, but is part of our deep-seated sense of a total developing pattern, imposing a responsibility on us for all that is to happen. Yet there is no power along with the responsibility: only in retrospect are we convinced that, if we had not done such-and-such a thing, the future would not have been doomed: we should have broken through the straitjacket of destiny. The more aware we are of what is happening in the world outside our immediate view, the more fully this kind of self-tormenting, this sense of total responsibility with a total absence of power, will impinge on our present moments.

Yet in another way this book insists on a multiplicity of futures. There is what will happen, regardless. There is also the future we can imagine, and this is as much part of our experience as the thing that must be. The Consul and Yvonne dream recurrently of a free life by the woods and waters of British Columbia, the more idyllic because of its remoteness alike from Mexico and from the more thronged stages of the world. Lowry himself, of course, spent much of his last years in British Columbia, and he is making no claim in his novel that even the natural splendour of Canada's west coast is a guarantee of contentment. But for Yvonne and the Consul it could appear to be, and the dream was part of their last day. The most precarious of dreams, of course: as Yvonne in Chapter IX sees in her mind the house above the beach she might share with her husband, she cannot fully concentrate on the simple image:

> Why was it though, that right in the centre of her brain, there should be a figure of a woman having hysterics, jerking like a puppet and banging her fists upon the ground?

And there is, too, the future beyond one's own death, which we have seen in personal terms in Chapter I, and which in Chapter XI, near the fact of death for the Consul and Yvonne, becomes extended beyond the time when this or that person of our 'now' can be remembered. The night sky will be observed so long as man lasts, and always the basic questions concerning purpose and cause will be raised. Then even Yvonne's infidelities and the Consul's drinking, even the battle of the Ebro and the things done in our time in the name of race or for gain's sake, will be forgotten: there will be only recurrence, of delight and fragile hope, of shame and loss, of questioning and failure to answer.

Meanwhile there is the present, charged full with memory and ambivalent response. An instant is indeed of copious content: under contemplation its extent can become enormous, as the stream-of-consciousness novelists have long made us know. Lowry has reinforced this by packing the one-day's sequence of Chapters II–XII with action and movement. It might appear that the intense activity could not have been borne by the three people concerned. The Consul begins the day hung-over and still drinking, Yvonne and Hugh have travelled overnight. Yet after her first sad encounter with her husband, she is seeing how her garden has become a wilderness, and then sets out with Hugh on their ride through the morning sunlight outside the town. Later they visit Laruelle, enter into the town's festivities on this day of solemnity, and then travel to Tomalín on the bus for the bullthrowing. Even then the journey is not over: there is the twilit road to Parián to be covered, where in the end the horse waits for Yvonne and the ravine for the Consul. When one thinks how little one does on most days, when one thinks of the dust and the heat and the nervous tension and the drinking, this day in a novel is frightening. Yet there are days that most of us have known, especially those days when tension has been at an extreme, which have strangely filled themselves with activity beyond normal endurance. In the result we are not wholly incredulous that the novel's people could have managed it, but we are conscious that such days are commonly days of crisis and shadow forth the characteristic depth of the existential moment. Yvonne in Chapter IX follows a boy's gaze into the sky as he looks for an airplane he can hear: 'she made it out for herself, a droning hyphen in abysmal blue.' That is an image for the momentary in its ungraspability, fugitiveness—in relation both to what we call 'now' in its strictest sense and to the lives we live in the whole context of time. But the hyphen drones, which suggests continuity, extension. The day, the individual's life, are also brief, but they take shape, they stretch profoundly by virtue of one's store of consciousness. What gives this extension is made up of personal relations and the nexus of cultural experiences. There is a lifetime in the moment, in the day of crisis. It is this that gives peculiar power to the novel or the drama that observes the Unity of Time, but this 'Unity' in its fullest employment is not one of mere restriction: it encompasses the all that a man knows.

Conrad and Lowry have used different technical methods. Paradoxically, the book called *Nostromo* is far less centred in the personal than the book called *Under the Volcano*. Lowry, with the apparently more restricted canvas, has also given the broader sense of historical event and geographical setting. By Conrad's standards, *Under the Volcano* is at times untidy, careless: not every detail is finely enough meshed with the growing pattern; there are times when the writer seems to argue with himself that everything, even a menu card, is of high interest because it exists. But both these novels are of major stature; both are remote from simple progression in their narrative; both explore the moment and see it within the total process; both are tragic writings which juxtapose the human consciousness with the irresistible operations of time and chance; both sharpen our sense of the immediate moment while insisting on that moment as only a point of intersection. This coheres with what we, the readers, know. For, like the modern novelist's material, our experience of past and future is being continuously shaped: our minds do the shaping now.

Chapter 6

ART AND THE
CONCEPT OF WILL

FROM the paintings of Paul Nash man is exiled. His landscapes are fairly trimmed, the fields are in good cultivation, the lines of trees are graceful and deliberate, but the agent who has disposed the scene has gone away. It is as if overnight the human race had destroyed itself but had left intact, for a moment, the evidence of its toil. This feature of Nash's work does not, I think, indicate that the artist romantically longs for the bliss of solitude, for a simple communion with the fields and the sky. Nor does it arise from that nostalgia we have previously remarked for a world untarnished by man's imprint.[1] Rather it seems to arise from a desire for singleness of vision. The artist is a maker, and at his best a maker of organisms, in which by definition every part exists by virtue of its function within a vital whole. Because a work of art is an attempt to master an area of personal experience, to bring into shape the haphazard materials that the external world offers to the artist's senses, the task will be easier if the materials are of the same order, if there is not one element among them with special claims, a special assertiveness. And man does notoriously assert himself, claiming from time to time a distinction from all other entities in the cosmos, the distinction of freedom of choice. This contrast between man and his environment was put neatly by Pope in 'The Univeral Prayer' that in 1738 he appended to his *Essay on Man*:

> Thou Great First Cause, least understood!
> Who all my Sense confin'd

---

[1] See above, ' "The Servants Will Do That For Us" ', p. 8.

To know but this, – that Thou art Good,
And that myself am blind:

Yet gave me, in this dark Estate,
To see the Good from Ill;
And binding Nature fast in Fate,
Left free the Human Will.[1]

Against the determined backcloth of nature, man alone dreams that he acts. The artist who includes man among the material that he uses in his construct must be prepared either to recognise a dichotomy in his material (and of such a kind that it will ultimately resist his efforts to organise it) or to deny the dichotomy and to insist that man's claims to freedom are erroneous. The second course is that followed by the Impressionist painters of the nineteenth century and by certain writers whom I shall be especially concerned with here. It is a course with powerful attractiveness and peculiar satisfactions, both for the artist and for his public. Thus we find Dryden addressing a fellow-playwright, the Earl of Orrery, and comparing his control over dramatic characters with God's control over men:

> Here is no chance, which you have not foreseen; all your heroes are more than your subjects, they are your creatures; and though they seem to move freely in all the sallies of their passions, yet you make destinies for them, which they cannot shun. They are moved (if I may dare to say so) like the rational creatures of the Almighty Poet, who walk at liberty, in their own opinion, because their fetters are invisible; when, indeed, the prison of their will is the more sure for being large; and instead of an absolute power over their actions, they have only a wretched desire of doing that which they cannot choose but do.[2]

As it happens, Dryden's claim for Orrery — a claim, clearly, that he would wish also to make for himself — is quite without substance: the dichotomy between man and his environment is nowhere more evident than in the serious drama of

[1] *The Poems of Alexander Pope*, ed. John Butt, London, 1963, p. 247.

[2] Dedication to *The Rival Ladies* (1664); *Essays of John Dryden,* ed. W. P. Ker, Oxford, 1926, I, 4.

the Restoration period. But the point of special interest is Dryden's hint at an artistic ideal. Its difficulty of consistent realisation is due partly to the artist's natural inclination to fall in love, both with real people and with his own characters. An almost inevitable consequence of falling in love is that one recognises the beloved as a free agent. The human beings in the world one contemplates, the figures in one's painting or novel or play, then acquire an assertiveness, refuse to be one with the milieu. And partly also the difficulty arises from the artist's irrepressible desire to modify his own environment: as a 'maker' this is not his concern, but he cannot usually keep entirely at bay the interest in the matter which he feels as a man. Zola, after a series of major novels in which he set out to present man's life as determined wholly by his heredity and environment, turned to a new vision in which the exceptional man was in a measure the controller of his life-pattern, the chooser of the right or wrong path. That these last novels of Zola represent a manifest decline of his powers is for the moment beside the point. Their importance lies in indicating the writer's human difficulty in adhering to the monistic view.

It will be evident that we are here concerned, not with the philosophic validity of free will or determinism, but rather with the problem confronting the artist. As a human being he is likely to have a sense, from moment to moment, that the disposition of event is at least partially in his control, and he is also likely to feel, when regarding as a whole either his own life or another's, that the general pattern has been ineluctable. This, we have seen, is how the reader feels in contemplating the action in *Nostromo*.[1] So in many great works of art one has a simultaneous consciousness of freedom of choice and rigid determinism. It appears to us as we see *Macbeth* that the hero is capable of choosing between his duty as subject, kinsman, host, and his murderous ambition. Yet simultaneously we are presented with the Weird Sisters who know every detail of future event, down to

[1] See above, 'The Shaping of Time', p. 100.

the movement of Birnam Wood and the fall of Macbeth in his fight with Macduff: looking back when the play is over, we cannot see any possibility other than what happens.[1] In the major English tragedies of the early seventeenth century this contradiction seems an inherent part of the pattern. One reason, perhaps, why these dramas are so moving is that they lack the tidiness of philosophic resolution. We have the sense in reading them that they objectify our double consciousness of the human life-pattern. They lead to a variety of interpretation largely because so many critics want to see in them a singleness of vision which is not to be found there.[2] They lack any rigorous systematisation, and the critic who imposes one on them contrives nothing but his own impoverishment. Yet the artist, as a maker, will long for full control of his material, and we as readers or spectators can derive a special pleasure from the work thus organised. This pleasure we can associate in particular with the plays of Chekhov and of John Ford, the paintings of the Impressionists, and the major novels of Zola. They depend alike on a denial of man's separateness from other elements in the cosmos. The denial, as we have seen, is precarious. It may be that we, as individuals, may consider the denial mistaken. Yet the artist in us will rejoice in what is here achieved: it is an image of the achievement of the Calvinist's God—a universe planned in all its variety, subdued wholly to its maker's hand.

There is a painting by the Canadian artist G. E. H. Macdonald, of the 'Group of Seven', hanging in the Faculty Club of the University of Toronto, which is commonly presented as a challenge to visitors to the University: 'Find the human figure there', they are urged. There is indeed almost at the centre of the canvas a small representative of humanity engaged in the portage of his canoe: the figure is so small, and so interwoven with the general design, that

[1]For a slightly fuller exploration of this matter, see 'The Dark Side of Macbeth', The Literary Half-yearly [Mysore], VIII (January and July 1967), 27–34.

[2] See above, 'Comedy in the Grand Style', p. 51.

sharp eyes are needed to find it. And when it is found, it does not stand against the natural scene but is merely part of it. There is no dichotomy, no human assertion. Its inclusion makes a point that Paul Nash refrains from.

The matter can be further illustrated, perhaps, by considering the development of stage-architecture in the last three and a half centuries. First we may note the basically separate natures of the Restoration and Elizabethan stages. In the theatre of Dryden's time there was a large fore-stage, on either side of which were two stage-doors, with boxes for spectators above. Behind the fore-stage there was a rudimentary proscenium-arch, and within the arch painted scenery was used to adorn and localise the action. The fore-stage was the place where the greater part of the play's action was presented. The actors normally entered and left the stage by means of the side-doors: they were very largely cut off from the pictorial background, which existed in a separate dimension, as it were, behind the proscenium-arch. At once we have the impression of human beings detached from the environment against which, rather than in which, they move. It is a kind of stage admirably disposed to suggest the dichotomy of man and his environment, and it is not surprising, therefore, that the characteristic serious drama of the Restoration is the Heroic Play, in which the hero is continually faced with a choice between duty and inclination, between honour and love, and admiration is demanded for the excellent man who exercises his volition in opting for the first alternative. The theatrical doctrine of Gordon Craig was firmly against the practice of setting down a three-dimensional actor against a two-dimensional background.[1] Because of its separation of the acting-area from the scene-area, the Restoration stage represented the extreme form of this juxtaposition of dissimilars, and for that very reason could effectively suggest the dichotomy between man and his environment that was posited in its most char-

[1] See, e.g., *On the Art of the Theatre*, London, reprinted 1957, implied *passim*.

acteristic plays. And, although the stage was gradually modified in the following two hundred years, the fore-stage shrinking and the action in general moving back within the proscenium-arch, the basic character of the eighteenth- and nineteenth-century theatre and drama is not far away from the Restoration. The actor is still a living man set against a patently contrived setting of painted board or canvas, the hero is still manifestly a free agent choosing honour—and love, too, if that is likewise honourable—and winning an audience's praise for his choice.

If we turn back to consider the Elizabethan stage, we see a radically different condition of things. There was no stress on pictorial background: if one were occasionally suggested, as with Henslowe's property 'the sittie of Rome',[1] it would not dominate the scene. The action took place mainly on a large platform jutting far out into the auditorium. This platform had rear-doors and practicable windows, and part of it was covered by a kind of roof supported by two pillars, respectively right-centre and left-centre of the stage. Behind there was usually, it appears, some kind of 'discoverable space', which could be used as an additional entrance, and somewhere above it an upper level for acting. The back wall of the stage may have looked rather like the façade of a large Elizabethan building, including doors and windows and with a pent-house projecting: this would give the stage something of an air of actuality, making it altogether closer to the world outside the theatre than the Restoration stage-picture could be.[2] The essential thing, however, is that we have here a three-dimensional stage on which the actor moves as in a real-life environment. He can hide behind the pillars, or climb them; he can open a window, or call to someone on the upper level as in actuality one might call to someone on a balcony or a city-wall; he

[1] *Henslowe's Diary*, ed. R. A. Foakes and R. T. Rickert, Cambridge, 1961, p. 319.
[2] This is of course a generalised picture: variations are likely enough to have been found in different theatres.

can shelter from sun or rain beneath the projecting roof. The audience, too, is almost all round him. Audience, actors, setting exist within the same order of being. Of course, the language of Elizabethan and Jacobean plays is characteristically rhetorical or poetic, and their pattern of event is often a good way from the normal current of life. The theatre gave a ceremonious presentation of life rather than even an impression of a replica. But the disposition of the stage was inimical to the sharp dichotomy of man and nature that we have seen in Restoration drama. Essentially the Elizabethan play mirrored our habitual confused responses to the human situation, while the Restoration play mirrored—much less successfully—an attempt to impose order on them.

To move on to the end of the nineteenth century and in particular to the work of Stanislavsky in the Moscow Art Theatre is to find ourselves with yet another set of conditions. Now actor and spectator are sharply separated. The actor lives on the stage in a setting that, as meticulously as possible, reproduces the appearance and character of the actual. Doors and windows and staircases are practicable, properties and furnishings are things as we know them in our own use. But here, as was not the case on the Elizabethan stage, the characters are subdued to the setting. They belong, not to the disordered world of the audience and the outside world, but to a world determined and disciplined. We see them consistently in the way that, at certain moments and reluctantly, we see ourselves, as particles having significance only within a totality of process. Kept rigorously apart from them by the proscenium-arch, we do not share with them our intimations or illusions of freedom. It will never be possible for the Three Sisters to move away in the direction of Moscow, Madame Ranevsky must lose her Cherry Orchard, because history has so willed it. That does not mean that Chekhov has no affection, or even a measure of approval, for his characters, but he is a man and artist more resolute than most, who is not impelled by liking to deduce the existence of capacity.

This was basically Ibsen's and Strindberg's theatre too: their plays have many characters who are fully articulate and rebellious, but the world they live in is their master. Peer Gynt and Brand, Rebecca West and Gregers Werle, do as they have to do, shut in and controlled by the conditions they resent. When Ibsen in *An Enemy of the People* or *The Lady from the Sea* talks of 'freedom', of acting on one's own responsibility, we may be moved yet remain ultimately incredulous.

In our own century this kind of vision has been more notably expressed in the cinema than in the theatre. There, indeed, even more than in Stanislavsky's playhouse, the world of the action is self-contained: actor and thing are alike images on a screen, and the performance is complete in every detail before the spectator is invited to observe it. Not surprisingly, the 'method'—a derivative from Stanislavsky—has been in recent years the hall-mark of the American screen-actor. He behaves, responding to his given nature and what is around him, rather than aims at suggesting the notion of a free agent. But it is especially in the major films from France in the last thirty years that we have been made aware of a consciously close study of humanity enmeshed in circumstance, as, remarkably, in Georges Rouquier's *Farrébique, ou les quatre Saisons*, where the cycle of human life in the French countryside was inextricably bound up with the march of the seasons, and in Jaques Becker's *Casque d'Or*, where the environment was urban but none the less compulsive. Within the playhouse, however, the most interesting developments have been away from Chekhov's naturalism. The fashion of theatrical speech has become more manifestly studied; the soliloquy and on occasion the use of verse have been reintroduced; the separateness of actor and spectator has been modified, the proscenium-arch being often seen as an enemy to be liquidated. And the drama has taken on something of the Restoration habit of thought. English verse-playwrights have joined hands with Sartre in claiming the possibility of an emancipated existence, differing from him only perhaps in the

115

character of the choice they recommend. This tendency to insist on the dichotomy of man and nature has been seen also in Ugo Betti's *The Queen and the Rebels*. Like the Restoration heroic dramas and much of the play-writing of Sartre, this play suffers from a too diagrammatic indication of the author's thesis, but in its very bareness it makes plain the basic character of its informing idea. The scene is an unnamed country in a time of revolution; the Queen is in flight, and strenuously pursued; by chance a woman of disordered life, beginning to age and to feel the increasing difficulties of her profession, comes upon the Queen and at first thinks to extort from her some object or information of value; acting on a kindly impulse, however, she lets the Queen escape, and this leads to her being herself mistaken for the Queen; she tries to establish her identity, but when that fails she finds a new and more glorious mode of existence in assuming aristocratic scorn and reviling her judges; when she is condemned, she could still save herself by betraying the Queen's friends, for the Queen has told her who they are; she refuses and, assisted it appears by divine prompting, she goes to execution with the assurance of having chosen the better path. The Sidney Carton parallel, though not to be pressed too far, is in some measure forced on us. We are back in the pre-naturalistic theatre, where man is seen as admirable when he does, as he can, rise above his environment, asserting his separateness, making his choice. The difficulty with this kind of writing, whether in Dryden's hands or in Eliot's, is that it too plainly sets itself against our normal responses to experience. Even if it is true that Nature is bound fast in fate while the human will is free, we do not sense the dichotomy as a thing naked and self-evident. Environment presses on us, and the burden of our inheritance: the scope of choice is at least restricted, and we are not altogether strangers in the cosmos. So, while we may feel that our common and confused perception of the human condition is presented in Elizabethan tragedy, we have the impression that Dryden or Eliot or Sartre gives us the dramatic equivalent of a *Reader's Digest* article—on

Descartes or on St. Thomas or on Sartre himself in his philosopher's gown.

In reaction from an existentialism too simply enunciated (and it is the simplification that is in question here), we may find to-day a more than usual pleasure in works of art at the extremest remove. For this reason the novels of Zola, after a long period of neglect and depreciation, have for some readers begun again to exercise a special authority. In his work we are a long way from notions of this or that kind of 'power'.

That indeed is not to say that Zola himself does not simplify. He takes over current scientific notions of the effects of heredity and environment, he posits (in the Rougon-Macquart series of novels) the hereditary burden that has been transmitted to his chief characters from Tante Dide and her husband and her lover, and then he places each character in an environment which exerts an influence upon him. The environment may be the great plain of La Beauce in *La Terre*, or the slums of Paris in *L'Assommoir*, or the world of Second Empire society in *La Curée*, or the northern coalfield of *Germinal*. Whatever it is, the environment is alive as much as the character, and is stronger than the character. Here, incidentally, his distinction from Hardy is apparent. Egdon Heath or Christminster can lead a man to despair, but Hardy nevertheless achieves a tragic effect by suggesting that his rebellious and defeated men and women are, in the ultimate scale, of a higher order than the environment that crushes them: he belongs, in fact, to the tragic and pre-naturalistic tradition. But in Zola the environment is not merely victorious: it is of the same order as man but with vaster and more powerful proportions. It has all the properties of life, for it not only destroys but is itself subject to change and is frightening in its transformations. There is a passage near the end of *L'Assommoir* in which Gervaise, near the end of her tether and her hope, goes out into the streets of Paris and suddenly realises the changes that have come upon the town in her life-time. The transformation of Paris under the Second Empire was never very far from Zola's mind. In a frenzy of 'enterprise', such as we may parallel in recent experience, the normal process of urban change was speeded up and made more evident. Man was not merely dominated by his environment, he was never sure what that environment would turn into. In *La Curée*, the second novel in the Rougon-Macquart series, the very centre of the book is the speculation in building-lots and building-projects under Napoleon III: the young wife of Aristide Rougon treads her path, without choice, towards incest and shame, while the city, it appears, writhes in the torment of its growth and men cluster where the money is. So, too, the pits of *Germinal* are precarious structures, ever on the point of physical collapse and abandonment, ever subject to economic change. Even the plain of La Beauce is not a stable countryside. Wordsworth could say that 'this frame of things'

> 'mid all revolution in the hopes
> And fears of men, doth still remain unchanged[1]

and elsewhere he intimates that, whatever cataclysm might come upon the earth, Nature would be essentially herself, still marked by 'composure' and 'kindlings like the morning'.[2] But for Zola Nature is never composed: it is part of the single compound to which man too belongs: it feeds him and exhausts him, and itself is fed and tormented and never still. In *La Terre* there is a character who dreams of the day when the whole plain of La Beauce will be fertilised by the sewers of Paris: he sees the establishment of a great cycle, the earth providing food to the people of the town, who in their turn provide from their bodies the means of further harvest. And the land can also decay, like the men who live on and from it, like the pits in *Germinal*: its cultivation is precarious, subject to the threats of war and of the flood of imports from the New World.

It is true that at the end of *Germinal* the sound of the men working in the pits below seems to Étienne Lantier like a

[1] *The Prelude*, Book XIV.
[2] *Ibid.*, Book V.

threat and a promise for the future. But it is a matter of an indeterminate future, a world not even hopefully Zola's. He impresses us with this ending, because it is in such manifest counterpoint to the world of the novel up to that point: he leaves us with a great question-mark. When D. H. Lawrence imitated this in the last chapter of *The Rainbow*, we could feel only an irrelevance. Zola had brought us into communion with the pit-workers, while for Lawrence, strangely enough, they were people outside the world of Ursula and her family which had been his consistent concern in his novel.

For men existing as part of the milieu that Zola presents to us, there is no possibility of greatness. This is where Zola chiefly simplifies, where he denudes us of part of our experience. In *War and Peace* Tolstoy was anxious to reduce the stature of the first Napoleon to that of a common man: for his great march through history the so-called leader was dependent, according to Tolstoy, on the casual chances of fortune. But in compensation Tolstoy's fictional characters exercise their will and, in a humanly imperfect fashion, put their impress on event. In his time it was, moreover, possible to believe that they could in large measure get things to rights in their immediate environment, whatever cataclysm had come on the world at large.[1] Zola, however, saw all men as made according to the pattern of his own smaller Napoleon: for him indeed all existence and every event were manifestations of a physical process. That is by no means to say that his attitude was nihilistic. He had a firm, and one may say an orthodox, sense of values. When in his last novels he turned to the positive inculcation of moral ideas, his preachments— though luridly dressed—were emphatically on the side of a simple and traditional virtue. And in the major novels there is no question where his sympathy and his approval lie. The curious gentleness and dignity that he gives to the girl Françoise in *La Terre*, the perceptions of the good that are

---

[1] In Piscator's dramatised version of the novel the difference from our own circumstances was made strongly evident. For further comment on this dramatisation, see below, 'The Dramatist's Experience', pp. 232, 236.

strong in Jean Macquart in the same novel and in Gervaise in *L'Assommoir*, in Étienne Lantier's impulse to social better-ment and his struggle for a reasonable control over the men and the pits of *Germinal*—all these coexist with the char-acters' complete subjection. For Zola, in his major writings, nothing is in a man's power to control, but man is not identical with man or thing with thing. We can rejoice or grieve that certain things, certain people, are as they are. Zola has been rightly and inevitably praised for his mastery of crowd-scenes and for his filling of a great canvas extended in both time and space, but within the plain of La Beauce and the whole history of the Rougon-Macquart family he showed that there was room for great diversity of human as of non-human life. He is, strangely, among the least monotonous of writers of long novels.

And he is helped in this by his comic as well as by his moral intuitions. It is frequently forgotten that Zola is sometimes at his most brilliant in scenes of broad but controlled comedy. The visit of the wedding-party to the Louvre in *L'Assommoir*, the more frightening comedy of the feast in the laundry in the same book, the many scenes in *La Terre* in-volving the rapscallion nicknamed Jésus-Christ (presented with an enjoyment and an absence of sentiment that may remind us of Shakespeare depicting Falstaff)—these things, existing in a setting of vice and squalor, remove Zola's world altogether from that of the tract and the white paper. He may deny human greatness and human responsibility, but human absurdity as well as human goodness lies within the range of his perception.

Among novelists I have chosen to refer principally to Zola here, but in a fuller treatment of the subject there would be much to say about the writings of Virginia Woolf and Ivy Compton-Burnett. Both of these use a narrow canvas, and we have the sense of the characters being hemmed in and controlled by their setting in place and time. For Virginia Woolf's men and women, more particularly her women, the consolation that could be extracted from life was the momentary perception, induced by love that went along with knowledge, of a pattern within the flux of things: her vision of the good was what we have so frequently seen to be that of the 'maker'. But the characters of Ivy Compton-Burnett have only the opportunity of playing with more or less vigour the parts for which they are cast. One may note in passing that a much greater variety in his approach to human life is possible for the novelist than for the dramatist who works fully within the playhouse of his time. The dramatist's work is largely conditioned by the theatrical architecture contemporary with him, and that is a thing necessarily slow to change. An uneasy assertion of the separateness of man and his environment was largely imposed on the dramatists of the eighteenth and nineteenth centuries, but in the novel we can find the eighteenth-century Sterne, with his gleeful insistence on man's puppet-like status, and the nineteenth-century Zola, the nineteenth-century Flaubert, coexisting with novelists—among whom are Richardson and Dostoyevsky—who see the possibility of human freedom while recognising fully the pressure of circumstance. Yet it is surely wrong to claim that the medium chosen always conditions the view of humanity that a writer offers: we may think of Büchner, a dramatist, asserting in the nineteenth century the tragic view, the dichotomy of man and nature, as fully as any novelist of his time. It was not a matter of theatre-architecture that led him to his characteristic dichotomy, but rather his essentially tragic idea of the human condition.

It is, of course, the complicated and contradictory vision of Shakespeare or Michael-Angelo or Dostoyevsky that accords best with our own intuitions when we are at our most vigilant and sensitive. It is to such men that we give the first place among artists. There is nothing in Zola so profoundly moving as that scene in *The Possessed* where Shatov's wife comes to his squalid lodging during the night—hysterical, reproachful, and near the point of travail. They have been long separated, and the child is not Shatov's. His ungainly anxiety to help, his almost schoolboyish manifestations of love, at length wear down her hostility, and when the child is born there is a radiance of affection on them as they plan

121

E'

the things they will henceforth share. As we read this passage, we already know that Shatov's brutal murder by his former friends is arranged for the next day. There is no hope of his escape, no hope that anyone in *The Possessed* can remake the world according to a pattern that he approves. We are reminded of the meeting of Lear and Cordelia near the end of their tragedy, where their doom is manifestly close but where they display an active will in the achievement of a transient good. It is difficult to explain why the characters of Dostoyevsky, like the major characters of Shakespeare, give us the fugitive—though only fugitive, as we have seen in the case of Macbeth—impression of having the freedom to choose, yet the sense of freedom does seem, for the momentary action, securely there. Despite the evident power of their environment, there is something daemonic in these men and women that defiantly claims a minimum of freedom. They have their 'spots of time' in which the shape of the world is for an instant at their disposal. That the land is at once laid waste again is no denial of their momentary power.

But there we are dealing with writing on the highest level, and we cannot expect dramatists and novelists to reach it often. Usually we find a simple assertion or a simple denial of man's freedom (or, on a lower level, a bland turning away from the issue), and the assertion may appear a grosser simplification than the denial. Within a determinist framework it is possible to have a rich and varied presentation of the human scene, while in heroic or sentimental writing the pattern of asserted will is likely to induce a numbing uniformity. Moreover, particularly for the dramatist, working within a restricted space, it is difficult to avoid making the act of choice appear too straightforwardly and lightly reached. So we may feel that Dryden's Almanzor too easily storms his way to honour, that Eliot's cocktail-drinkers and confidential clerks step too simply into martyrdom and God's ministry; while the opposite simplification—the displaying of man denuded of will, of the capacity which, if it exists, is the supreme distinction—seems at least a little closer to our general perception of the nature of things.

Chapter 7

CATHARSIS IN ENGLISH
RENAISSANCE DRAMA

THERE is fairly general agreement on the historical
origins of the trouble about *catharsis*. Plato in *The Republic*
condemned poetry partly because it called into play certain
emotions, including pity and fear, which prevented a man
from thinking rationally.[1] Aristotle in *The Poetics* did not
deny that pity and fear were strongly experienced by those
who saw or read a tragedy—he could not, indeed, deny what
was a self-evident truth—but in rebutting Plato's argument
he took a hint from the current use of music in the treatment
of mental disorder. It was the practice to increase the degree
of violence in the patient by the playing of musical instruments, until a state of exhaustion set it.[2] As recently as
twenty-five years ago in the Arabic speaking world, and
doubtless it survives to-day, the practice of *ez-zar* was basically what Aristotle knew in Hellas: only it had become semi-
rationalised in associating particular tunes with particular
spirits, and the patient, who was regarded as possessed, was
abandoned by his demon when the right tune was played.[3]
Of course, it is a matter of dispute whether Aristotle believed
that *catharsis* worked as a means of evacuating pity and fear
or whether his notion was that through tragedy those emotions were themselves purged of the dross in them—so that

[1] Book III.

[2] S. H. Butcher, *Aristotle's Theory of Poetry and Fine Art*, fourth edition,
London, 1932, pp. 248–9.

[3] Cf. *The Encylopaedia of Islam*, Leyden and London, 1934, IV, 1217.
For a late nineteenth-century account of the survival of the rite, see
André Gide, *Journals 1889–1949*, translated, selected and edited by
Justin O'Brien, Harmondsworth, 1967, pp. 50–1.

they were no longer the handicaps to good living that Plato found them, but functioned, along with his reasoning powers, as part of a man's necessary equipment for living. It is the second of these interpretations that seems to be the dominant one to-day, and it has been put clearly and eloquently by Humphry House:

> A tragedy rouses the emotions from potentiality to activity by worthy and adequate stimuli; it controls them by directing them to the right objects in the right way; and exercises them within the limits of the play as the emotions of the good man would be exercised. When they subside to potentiality again after the play is over it is a more 'trained' potentiality than before. This is what Aristotle calls κάθαρσις. Our responses are brought nearer to those of the good and wise man.[1]

What Aristotle meant is for Aristotelians to argue over. What we must rather be concerned with is whether either of these or any other available notion of *catharsis* has any relevance to the writing and the witnessing of tragedy. What I shall suggest here is that what underlies Aristotelian *catharsis* has been of major importance in tragic writing, even when Aristotle has not been much or at all in the dramatist's mind, but that the idea itself has been frequently and seriously damaging to tragic theory. Of course we all know how the word can be thoughtlessly used as a vague gesture of approval. F. L Lucas made merry with this when he said:

> Indeed the *Catharsis* does yeoman service still: if we dislike A's poem or B's play, we need only say impressively that they fail to produce 'the right cathartic effect'. It is much simpler than giving reasons.[2]

What is more important is that the notion of a cathartic effect as being at the root of Greek and Renaissance tragedies has for some contemporary minds made those major writings seem substantially irrelevant to the world in which we now live. Maynard Mack has recently commented, with charitable

[1] *Aristotle's Poetics*, London, 1956, pp. 109–10.

[2] *Tragedy in relation to Aristotle's Poetics*, London, 1927, p. 23.

disapproval, on Charles Marovitz's account of his and Peter Brook's aims in their production of *King Lear*. 'One problem with *Lear*', said Marovitz, 'is that like all great tragedies it produces a catharsis. The audience leaves the theatre shaken but assured.' Now Brook and Marovitz wanted them shaken but not assured. So they decided to eliminate from the play anything that might hint at consolation, any suggestion that things might not be wholly wrong, any easing of our situation through the establishment of sympathy between a character and ourselves. So Lear's bad behaviour must be very bad; Cornwall's servants must not turn against their master when Gloucester is blinded; Gloucester himself must be pushed aside by scene-shifters so that, even in his moment of sharpest suffering, he is treated as a thing of no account; and at the end of the play a rumbling of the storm must be heard again, lest we should think that anything had been firmly brought into a state of order.[1] On the last matter the production would appear to have taken a hint from E. M. Forster's tragic interpretation of Beethoven's Fifth Symphony in *Howards End*, where the perceptive listener, we are told, will realise that the goblins, having come back once before, may come back again.[2] I should indeed have thought that this was as plain for Shakespeare as for Beethoven, that we do not need a return of the storm at the end of *Lear* any more than we need goblin-notes at the end of the Fifth: the implication should be strong enough in each case.[3] There is a remarkable irony in the fact that Shakespeare's text was thought too terrible for the stage from the late seventeenth to the mid-nineteenth century, and now for some it appears not terrible enough. We must switch the emphasis, we must omit the so-called cathartic touches, just as Nahum Tate in 1681 switched the emphasis in another way

[1] Maynard Mack, *King Lear in Our Time*, Berkeley and Los Angeles, 1965, pp. 30–2, 39–40. Marovitz's account appeared in '*Lear* Log', *Encore*, X (1963), 22. For further comment on the production, see Alfred Harbage, *Conceptions of Shakespeare*, Cambridge, Mass., 1966, pp. 71–5.

[2] Chapter V.

[3] See above, 'When Writing Becomes Absurd', p. 78.

and omitted the things that worked against 'poetical justice'. But the parallel is inexact: Tate altered the play to make it fit with a critical notion that he thought Shakespeare had wrongly neglected and that Thomas Rymer had made explicit in his *The Tragedies of the Last Age* in 1677; Brook and Marovitz altered it because they felt it did adhere to a critical notion which they found repugnant or at least inappropriate for our time. I doubt whether they would have arrived at this point if their notion of *catharsis* had not been over-simple.

It would be strange if Aristotle, merely to rebut Plato's argument, had quite arbitrarily imported into his theory of tragedy a notion that he found conveniently at hand in the sister-art of music. Even the strongly musical character of ancient tragedy would not justify such a glib transference from one field to another. Moreover, as D. M. Hill has pointed out, in music the effect worked on the sick: others, as Aristotle says in *The Politics*, may take an 'innocent delight' in the melodies, but since they are in good health there is no therapeutic effect on them.[1] So what Aristotle can hardly have had in mind was a general curative process in tragedy, at least in the same fashion as operated, but merely for sufferers, with music. Yet he could observe what happened in a theatre, and we are not likely to be straining at probability if we assume that, in Athens in his time as in London or Paris or New York in ours, an audience which had just witnessed a tragedy did behave in a special way. This behaviour may perhaps be best understood by using the method of exclusion: the spectator of tragedy does not think of a programme of action, turning from the play to see that this terrible thing does not happen again; he does not, for the moment, want to storm a Bastille or throw up barricades or join a party. This indeed we have seen is what we should not expect, typically, from the contemplation of any work of art.[2] But neither does the spectator feel that all is now well or even easily toler-

---

[1] '"Catharsis": An Excision from the Dictionary of Critical Terms', *Essays in Criticism*, VIII (1958), 113–19; reprinted in L. Michel and R. B. Sewall (edd.), *Tragedy: Modern Essays in Criticism*, Englewood Cliffs, 1963.

[2] See above, 'A School of Criticism', p. 40.

able, either within the *polis* or within the individual mind. He is passive in relation to any particularised plan of action, but is not acquiescent in the nature of things. At the most he has a sense of respite: a terrible vision has been put before him, and the sequence of terrors has for a moment halted. As Forster, we have seen, put it, the goblins can come back; there will be more Iagos to seduce more Othellos; a deep involvement with another person will again lead to murder, as it did for Medea when she killed her children, for Ferdinand of Calabria when he gave his sister the Duchess of Malfi to a ritualised death, for Giovanni in *'Tis Pity she's a Whore* when he tore out Annabella's heart. But the performance is over; we hold it in memory; we recognise its general truth, and in doing so render the spectacle more abstract; we move back from the witnessing of violence and return to normal and less intense relations. Because indeed we have a sense of respite, we are in a better position to assimilate the tragic doctrine. We are quiet as we carry these things in our heart. It is true that not all so-called tragedies in Aristotle's time ended like this. The *Oresteia* and apparently the *Prometheia*, when viewed as wholes, as they would be in a day-long performance, offered notions of civic and cosmic achievement: the court of the Areopagus would enable men to come nearer in practice to their highest ideal of equity; the reconciliation of Zeus and Prometheus pointed towards the finding of a place, within a total coherence, both for traditional law and for the impulse to rebellion and adventure and a lawless charity.[1] Moreover, some so-called tragedies moved away from the terrible into a vision of human littleness and dependence. The *Alcestis* has an irony of the comic kind: we take away a sense of discomfort, not a mind over-charged with a spectacle of anguish that tragedy proper gives to us. In *The Poetics* Aristotle in his usual way was concerned with description rather than with prescription, so he did not say this play is tragedy and not that: he accepted the practices

---

[1] Cf. H. D. F. Kitto, *Form and Meaning in Drama*, London, 1956, pp. 69–72, 81–2.

and the terminology of the poets and their theatre. Yet when he talks specifically of a tragic effect we can be reasonably sure that it was with reference to such plays as *Oedipus* (both *Tyrannus* and *Coloneus*), *Antigone*, *Hippolytus*, the *Electras* of Sophocles and Euripides, and all those others which confront us with desolation: 'none can be called happy until that day when he carries His happiness down to the grave in peace'[1] is how *Oedipus Tyrannus* ends, and this is a long way from reconciling us with life. But of course tragedy does not give us only desolation: it gives us an image of life more abundantly lived than we ordinarily know it. We feel more alive for seeing Oedipus or Antigone in front of us, and the intensity of the experience both balances the desolation and makes it more acute. The element of awe in the experience, the sense of human magnitude, also helps to make us quiet, for we do not easily find words to describe these people and their actions. The dramatist may indeed directly embody in his play's ending our sense of being at a loss for words, as Shakespeare did in the last lines of *Lear*:

> The weight of this sad time we must obey;
> Speak what we feel, not what we ought to say.
> The oldest hath borne most; we that are young
> Shall never see so much nor live so long.
>
> (V. iii. 323–6)

Edgar[2] here voices our inability to find words for 'what we ought to say', and then he falls into a silence of admiration for the prime figure in the tragedy, a silence at the unspeakable things that figure has known. Like us, Edgar grows quiet. In the Royal Shakespeare Theatre's production of the play in 1968 there was an eloquent pause before the lines were spoken: the survivors looked at one another to see who could find something tolerable to say; there was a sense

[1] *Sophocles: The Theban Plays*, tr. E. F. Watling, Harmondsworth, 1947.

[2] The lines are Albany's in the Quarto, but I think less appropriately: Edgar through the play is the more neutral figure, the one more fitting therefore for a last semi-choric—personal yet choric, one might phrase it, as so often in Shakespeare—utterance.

of relief when Edgar took up the burden, and the very inadequacy of his words made them more appropriate and more moving. The seeing of things like this in the theatre might indeed have led Aristotle to talk of *catharsis*. But we are emptied of words, not of passion: indeed, we are more passionate than before. And I see no reason to think that we shall behave more discriminatingly in future in the direction of our pity and fear. Life will not be easier for us through what we have seen. But at least we have seen. We know more.

There is a different matter, however, for us to reckon with, a matter given special and illuminating attention in John Holloway's book on Shakespearian tragedy, *The Story of the Night*.[1] There it is pointed out that in Shakespeare's tragedies—and most obviously in *Lear*, *Macbeth*, *Coriolanus*, *Timon*—the hero is simultaneously exalted and increasingly isolated; that the hero comes to accept his ultimate destruction, finding a rightness in it; that this destruction increases a sense of cohesiveness within the society to which the hero both belongs and does not belong. We can see it in these Shakespearian plays, certainly, but it is also evidently true, in a greater or lesser degree, of some major Greek tragedies. Thebes is sick, and one man must die for the people, one man must carry their collective guilt and be driven out or at least debased and removed from his high office. It is appropriate that it is the king, the living symbol of state, who thus suffers, who thus by his willing death or abdication purges the guilt from the minds of the community which watches the act done. The point was given special prominence in Sartre's *Les Mouches*:

> *Orestes.* ... As for your sins and your remorse, your night-fears, and the crime Aegistheus committed—all are mine, I take them all upon me. Fear your Dead no longer; they are *my* Dead. And, see, your faithful flies have left you, and come to me. But have no fear, people of Argos. I shall not sit on my victim's throne or take the sceptre in my blood-stained hands. A god offered it to me, and I said 'No'. I wish to be a king without a

[1] London, 1961; see especially Chapters V, VIII.

kingdom, without subjects. Farewell, my people. Try to reshape your lives. All here is new, all must begin anew. And for me, too, a new life is beginning. A strange life . . .[1]

And of course to some extent we still feel this outside the theatre, however accidental the choice of a victim who can be seen to stand for the community. This can explain the communal rage recurrently felt to-day against the public figure who can be associated with failings widespread in our society.

It is the way of tragedy to present a sick land, sometimes through the guilt of the governor, as when Macbeth tells the Doctor:

> If thou couldst, doctor, cast
> The water of my land, find her disease,
> And purge it to a sound and pristine health,
> I would applaud thee to the very echo,
> That should applaud again.   (V. iii. 50–4).

But sometimes the community is sick without reference to an act of the governor, as Antonio suggests at the beginning of *The Duchess of Malfi*:

> a prince's court
> Is like a common fountain, whence should flow
> Pure silver drops in general: but if't chance
> Some curs'd example poison 't near the head,
> *Death, and diseases through the whole land spread.*
>   (I. i. 11–15)[2]

This helps us to find the Duchess as the necessary and purging victim when she comes to her ceremonial death in Act IV. If we remember the religious atmosphere of a tragic performance in Athens, we are likely to recognise that in such a context the notion of a scapegoat would be particularly strong. Agamemnon dies for a shared Greek guilt in the

[1] Jean-Paul Sartre, *The Flies and In Camera*, tr. Stuart Gilbert, London, 1946, p. 102.

[2] *The Duchess of Malfi*, ed. J. R. Brown (The Revels Plays), London, 1964.

pillaging of Troy and the recurrent brutalities that marked the expedition; the unintentional guilt of Oedipus has become the guilt of the town which made him king, and he must suffer that the plague may be lifted; Antigone must die because, in her defiant innocence and her defiant breaking of Creon's law, she has rebelled as we dare not rebel: paradoxically, therefore, she atones for our craven hearts. And with Elizabethans or ourselves seeing tragedy enacted, there is again a sense that a man has purged a corrupt society through his death. The state of purgation will not last long. There must be recurrent victims (yearly victims in some societies, as Frazer pointed out[1] and as Mary Renault has movingly reminded us in *The King Must Die*[2]), but for the moment our guilt is not with us. 'Christ is risen', we may exclaim every year, and at the following Good Friday he must die again. Eliot in *The Waste Land* told us that it was for a very brief time that the rains came and the earth was fertile. It is impossible to say how conscious the earliest tragic writers were that they were staging a scapegoat ritual, but there can be little doubt that that in part is what tragedy is, and that the feeling of respite it brings is in part due to the strongly remaining sense that the ritual still works. Let there be a sacrifice: we shall be appalled, but it will have carried away the immediate burden of our sins. So in *Les Mouches*, Sartre's already referred to rehandling of the Orestes story, there is emphasis on how the people of Argos had a 'grand time', a 'real gala night', when they went to bed after the killing of Agamemnon.[3] Though neither he nor Oedipus nor Lear is our king, though the particular community represented is far away (as Argos and Thebes must have seemed for Athenians in the fifth century B.C.), in watching their deaths or other humiliations we too are for a time purged and therefore fortified.

But of course it is not so simple as that, and John Hollo-

---

[1] J. G. Frazer, *The Golden Bough*, abridged edition, London, 1947, p. 575.
[2] London, 1958; see especially Book II, Chapter I.
[3] *Ed. cit.*, p. 14.

way has I think stopped short in his relating of tragic drama to primitive rite. Tragedy is the product of a highly self-conscious society, of, for example, fifth-century Athens, seventeenth-century London or Paris, and such societies are not to be equated with those that get a straightforward relief from a scapegoat ritual. The ritual remains, in a more complex form, but the effect too has become complex and contradictory. That the king has died gives us relief, but we are simultaneously revolted at the need for him to die. The whole system of things comes under our question, as it does in *Doctor Faustus* and *Lear* and *The Duchess of Malfi*. Rationally we know that we have not been purged, however strong the vestigial sense that a purging has taken place. And even if we were thus relieved, we should resent it. We do resent it, for has not a solemn ceremonial been powerfully suggesting that the old rite, the old demand from on high, the old psychological need from within, are still with us? So the effect of tragedy is double once again: the peace of vicarious atonement, the resentment of things within and without that have made us want it. But here indeed we can say the effect is a treble one, for we feel shame at our need as well as rebellion against its being implanted within us. Previously we have seen the effect of tragedy as desolation balanced against a pride in the hero's capacity to die more nobly than we can hope to. Now it can be seen as giving us both relief and resentment and shame. In taking away the sense of collective guilt, it has made us recognise a special and civilised guilt in our very acquiescence in the rite.

It follows that no true tragedy gives us simple relief or can possibly be a straightforward communication of doctrine.[1] When those things do happen, as with Addison's *Cato* or Milton's *Samson Agonistes*, we have what I. A. Richards has rightly called 'pseudo-tragedy', though he does not refer explicitly to those examples.[2] Before turning to the Jacobeans and the Elizabethans it will be profitable to stay

---

[1] See above, 'A School of Criticism', p. 40.

[2] *Principles of Literary Criticism*, fifth edition, London, 1934, p. 247.

for a moment with their great castigator who in *L'Allegro* had shown delight in Jonson and Shakespeare, who had contributed verses to the 1632 Folio, but who in 1671 felt it necessary to defend the tragic kind because it had fallen so low in his time:

> *Gregory Nazianzen* a Father of the Church, thought it not unbeseeming the sanctity of his person to write a Tragedy, which he entitl'd *Christ suffering*. This is mention'd to vindicate Tragedy from the small esteem, or rather infamy, which in the account of many it undergoes at this day with other common Interludes; hap'ning through the Poets error of intermixing Comic stuff with Tragic sadness and gravity; or introducing trivial and vulgar persons, which by all judicious hath bin counted absurd; and brought in without discretion, corruptly to gratifie the people.[1]

When Milton says 'at this day', he cannot have anything specifically Restoration in mind: his scornful words apply easily to drama before the Civil War, but will not fit the heroic plays which Orrery and Dryden had begun to write in the decade before the publication of *Samson Agonistes*. In his own play, which was not to be tainted by performance, Milton would show how tragedy in its pure and ancient form could still be written but within the framework of Christian belief. And in the first sentence of his preface he refers to *catharsis*: tragedy, we are told, is

> said by *Aristotle* to be of power by raising pity and fear, or terror, to purge the mind of those and such like passions, that is to temper and reduce them to just measure with a kind of delight, stirr'd up by reading or seeing those passions well imitated.

Milton goes on to illustrate the process from the physical world:

> for so in Physic things of melancholic hue and quality are us'd against melancholy, sowr against sowr, salt to remove salt humours.

[1] Quotations from *Samson Agonistes* are from *The Poetical Works of John Milton*, ed. H. C. Beeching, London, 1922.

He does not say that pity and fear will be eliminated but reduced 'to just measure', thus perhaps taking a half-way position between those who interpret *catharsis* as total evacuation and those for whom it is a better ordering of the passions themselves. In the play that follows, he seems closer to the second of these views. We begin in a corrupt society, the Gaza of God's enemies the Philistines, with the Israelites disgraced and their champion not only eyeless and bound but aware of his own corruption. We are made to see him conquer the desire to withdraw into peaceful retirement with his father Manoa, the desire for the private comfort of Dalila's bed, and then to see him becoming aware that God is giving him another chance—a chance to die for his people and thus to be accepted again as God's and their champion. He has been personally guilty in the marriage with Dalila and in his indiscreet babbling with her; he is also, as the great Israelite, the image of tribal being and tribal guilt. It is therefore fully appropriate that in redeeming himself he frees his people from immediate bondage and symbolises the total redemption that will later, through another, be offered to them. Through his personal *felix culpa*, he becomes especially fitted to be a type of Christ, the Second Adam. When at the end the sacrifice has been made, the hero's father does not suggest we have no cause for lamentation, but rather that we have greater cause for rejoicing and that at this moment it would be inappropriate to give licence to grief:

> Come, come, no time for lamentation now,
> Nor much more cause, *Samson* hath quit himself
> Like *Samson*, and heroicly hath finish'd
> A life Heroic, on his Enemies.
> Fully reveng'd, hath left them years of mourning,
> And lamentation to the Sons of *Caphtor*
> Throgh all *Philistian* bounds.
>
> (ll. 1708–14)

And the famous last words of all invoke the greater knowledge of God's ways, the 'peace' and 'consolation', the 'calm' and the passionlessness, that the great event has for the time being induced:

> His servants he with new acquist
> Of true experience from this great event
> With peace and consolation hath dismist,
> And calm of mind all passion spent.
>
> (ll. 1755–8)

It should be clear that 'all passion spent' does not imply an absence of emotion but a condition in which there is no disturbance of the understanding through passion—that is, through untoward or disproportionate emotion. When Ferdinand in *The Tempest* declares that the 'white cold virgin snow' from his heart abates the ardour of his liver, he does not imply that his attitude to Miranda is untinged by emotion, but rather that the emotion is entirely proper. Thus in *Samson* we have an attempt to write tragedy in which the idea of *catharsis* is central to the writer's purpose. It does not fully work for us to-day, partly on humanitarian grounds (for we cannot see why there should be such simple rejoicing at the slaughter of a vast number of Philistines), partly because the myth is too directly presented (for we have difficulty in accepting mass-slaughter as an image of atonement), partly because many of us can no longer find an acceptable god in the one who is celebrated here. What is rigorously avoided in this play, moreover, is any substantial kind of ambivalence. The final chorus admits that 'we oft doubt' and that often God 'seems to hide his face', but at once asserts that what we have just seen is a glorious witness of his power and wisdom. Even so, the reader is not told to take himself to simple rejoicing: rather, he has something to hold quietly in his heart, and he is to find room for a modest lamenting at Samson's loss when the time has grown ripe for it. So we have the final quietness and contemplation and the sense of relief through the sacrifice of a man for the people—all things that we have found characteristic of tragedy. But there is nothing of the resentment that a scapegoat should be needed by us or by the cosmic powers, nothing of the desolation that went along with the pride in the tragedies we have previously noted. *Samson Agonistes* is a simpler thing than *Oedipus* or *Lear*: it represents an ingenious wedding of the *catharsis*-formula to Christian myth. Perhaps alone among tragic writings, it insists that the reader should think in terms of *catharsis*. Elsewhere in seventeenth-century tragedy, we shall see, the notion is far less overt and the total effect of a play becomes distorted if the reader tries to force it into Milton's pattern.

Certainly we can find formal tributes to the dead hero in the plays written for the seventeenth-century theatre, especially in Shakespeare's, but always with the dominant sense that a gap in nature has been left by his death. Some non-verbal gesture of lamentation is often added to the words, as a dead march is sounded, for example, in *Lear*, or the soldiers shoot at Elsinore. Shakespeare is particularly ingenious in varying the pattern of the tribute, having it spoken by the hero's successor in *Hamlet*, by his victorious adversary in *Julius Caesar* and *Coriolanus*, by his bereft mistress in *Antony and Cleopatra* (and she, characteristically, does it twice over, at the end of Act IV and in talk with Dolabella in Act V), and by the hero himself in *Othello*. In this last instance the hero refers to his own weeping at his situation, in remarkable contradistinction to the forbidding of our tears in *Samson*:

> Then must you speak
> Of one that lov'd not wisely, but too well;
> Of one not easily jealous, but, being wrought,
> Perplexed in the extreme; of one whose hand,
> Like the base Indian, threw a pearl away
> Richer than all his tribe; of one whose subdu'd eyes,
> Albeit unused to the melting mood,
> Drops tears as fast as the Arabian trees
> Their med'cinable gum.   (V. ii. 343–51)

Alone of Shakespeare's tragedies, *Macbeth* contains no such tribute: the hero and his wife are merely 'This dead butcher and his fiend-like queen', and in that play Shakespeare lets the ambiguities remain indeed always implicit. Yet, to labour neither that nor other complications for the moment,[1] we are likely to agree that in each of these instances, and in the tragedies of Chapman and Webster and Jonson, we feel that a figure of authority and high vitality has gone, that a kind of normality, though an impoverished normality, has been returned to through his death. In the face of the individual's destruction the cosmos is darkened, but the small world of society can breathe more easily. On the one side there is grief, there is puzzlement, there is the sense of an ultimate silence in which our questionings become vain; on the other there is relief as the hero seems to have died that the current of life may run more quietly.

The second part of this effect is reinforced by the omni-presence in early seventeenth-century English drama of images of corruption and disease. And we should note that this is true of comedy as well as tragedy in these years. We have already observed it in *Macbeth* and *The Duchess of Malfi*, and it is easy enough to add to these examples. 'Something is rotten in the state of Denmark', cries Marcellus when he has seen the Ghost indicate a readiness to speak with Hamlet. And thus the disguised Duke in *Measure for Measure* refers to the condition of his city:

> My business in this state
> Made me a looker-on here in Vienna,
> Where I have seen corruption boil and bubble
> Till it o'errun the stew.   (V. i. 318–21)

Sometimes the image applies to a sickness within the in-dividual person, as when Lear cries to Goneril:

> But yet thou art my flesh, my blood, my daughter;
> Or rather a disease that's in my flesh,
> Which I must needs call mine; thou art a boil,
> A plague-sore, or embossed carbuncle
> In my corrupted blood.   (II. iv. 224–8)

---

[1] But see above, 'When Writing Becomes Absurd', p. 82.

Or when Beatrice-Joanna in *The Changeling*, at last conscious of her nature, tells her father:

> Oh come not near me, sir, I shall defile you:
> I am that of your blood was taken from you
> For your better health; look no more upon't,
> But cast it to the ground regardlessly:
> Let the common sewer take it from distinction.
>
> (V. iii. 149–53)[1]

With such things brought to our mind, we are bound to be led to the feeling that blood-letting is needed, just as the men of the time practised it in physic, just as they imagined that war was recurrently necessary for the purging of the ills that a long peace brought on society. Indeed in certain exceptional plays, like *Measure for Measure*, Marston's *The Malcontent* and Ford's *The Fancies Chaste and Noble*, where there is the stress on corruption but no blood-letting takes place, we are likely to feel as great an uneasiness as in the ending of a tragedy. Though on different levels of achievement, these three plays are alike in representing a special kind of dramatic writing which we may call the comedy of frustrated expectation, where purging has been made to seem necessary but where the knife has drawn back. We need, I think, at this point to consider whether *catharsis* has some relation to comedy.

Aristotle, of course, is brief on comedy and makes no mention of *catharsis* in connection with it. If, however, there is any substance in what I have laid out as a series of explanations of how he was led to the idea of a tragic *catharsis*—the analogue of music as therapy; the sense of the final pause for contemplation and assimilation; the survival into tragedy of purgation-feelings associated with a scapegoat ritual—then I think we may be justified in looking for related phenomena in comedy. At this point, on the other hand, I should emphasise that we may be entirely wrong in assuming such ideas as inherent in Aristotle. Gerald F. Else in *Aristotle's*

[1]*The Changeling*, ed. N. W. Bawcutt (The Revels Plays), London, 1958.

Poetics: The Argument[1] has urged that what Aristotle meant by *catharsis* was the purification of the tragic event, so that an action which would be intolerable to contemplate in ordinary circumstances, such as parricide or incest, is rendered tolerable by the special circumstances devised by the dramatist. Despite what he has done, Oedipus is demonstrated to be a man of integrity and so he can be legitimately pitied. Consequently, says Else, *catharsis* cannot exist in comedy, where there is no 'deed of horror' to be cleansed. But, if that after all is what Aristotle meant, we can say at once that it has not much relevance to Greek tragedy (as indeed Else recognises) and perhaps no relevance at all to subsequent tragedy. Medea's killing of her children, Electra's urging of her brother to revenge (in both Sophocles and Euripides)—these are not purified in our eyes. And in later times when dramatists have had Aristotle in mind (most obviously, of course, Milton), the cathartic effect is understood in relation to our responses, not to the deed done. Milton would not feel it necessary to excuse or purify the slaughtering of God's enemies. And whether or not the Jacobeans were much aware of Aristotle, what they did seems more in conformity with the explanation of *catharsis* I have suggested than with what Else believes was intended in *The Poetics*. Paradoxically we may urge that, if Else is right, a non-Aristotelian *catharsis* has had more validity than the one that he himself devised.

So let us, presumptuously, set Aristotle aside for the moment, and also set aside the analogue from music, which in any event I think was no more than a convenient help for him in rationalising the experience of a final calm. And let us ask whether comedy gives us occasion for a sense of a scapegoat ritual and for the sense of a pause for contemplation and assimilation. I should like here to refer to some passages in David Storey's novel *Radcliffe*.[2] The central

[1] Cambridge, Mass., 1957. The particularly relevant passages are on pp. 221–32, 423–5, 436–47, which are reprinted in Michel and Sewall, *op. cit.*

[2] London, 1963.

character is taken to see a comedian appear in the crudest kind of solo entertainment: to his antics and feeble bawdy the near-moronic audience responds by seeing him as their inferior, delighting in thus finding themselves elevated, momentarily secure. The comedian's very name, for professional purposes, is a badge of humiliation: 'Gormless Gordon the gump from Gorseforth'. While the audience is delighted, Radcliffe is appalled and, when he meets the comedian afterwards, accuses him of pandering to them. The comedian's reply is:

> 'Don't you see? It's *their* humour, not mine. But I'm not apologizing for them. Nor for myself. At some other club I do something entirely different. Just singing straight ballads. But I want you to understand: *this is all they've got.*'[1]

These people can laugh, can feel secure, in finding someone they can despise. The comedian offers himself for their contempt. So does Archie Rice in John Osborne's *The Entertainer*, and like Gormless Gordon he ultimately turns on the audience, telling them they are no better, or indeed worse, than he is. In both these instances we are given a picture of a sacrificial victim no longer acquiescent. With such figures comedy functions on its most obvious level (the comedian doing a solo act), but even in less direct ways it can expose human beings to lasting scorn. Pistol and Ananias and Bessus and Ajax, whose abasement may appear less naked because of its complex setting, would in isolation be merely instruments for our relief from insecurity.

Comedy, however, frequently works more gently. Tragedy kills its victim, or at least puts him outside the pale; comedy has a way of rehabilitating him. I have already drawn attention to the fact that English comedies as well as tragedies around the year 1600 exhibit pictures of corruption. In most comedies the corruption is merely that of folly. It may be the affectation of grief in Olivia, of love-sickness in Orsino; it may be the elderly lechery of Falstaff, the jealousy of Ford,

---

[1] p. 164 of the Penguin reprint.

the eye to the main chance in the parents of Anne Page; it may be the melodramatically presented tyranny of Duke Frederick in *As You Like It*. These things can be put right with the help of a little luck, a little ingenuity. So in *Every Man in his Humour* Old Knowell and Downright and Kitely can be cured of their excessive humours through the logic of event and the shrewdness of Justice Clement. Comedy, in fact, is in one of its aspects able to bring eccentricity to school, to make it conform with society's conception of how men should behave. We are made to see how folly manifests itself, and are reinforced in our view of the wisdom of the social norm. The effect is not to cure us of our follies, but rather to give us the sense that folly, having been objectified and dramatically dealt with, is no threat to our well-being. Lessing in his *Hamburgische Dramaturgie* put it, perhaps too gently, in this way:

> Granted that Molière's Miser never cured a miser, or Regnard's Gambler a gambler; admitted that laughter could never cure these fools; the worse for them, but not for comedy. Comedy, if it cannot cure desperate diseases, is satisfied with fortifying the healthy in their health. The Miser is instructive also to the generous man; and he who never plays may yet be edified by the Gambler.[1]

What happens is that we see men on the stage committing follies we might commit or have committed; we see them being dealt with and cured; vicariously we re-enter society with them; they have publicly suffered for what we, either actually or potentially, have committed. We are glad to have got off so lightly.

But of course we have not got off. Are we any more immune from folly for seeing comedy expose it? Lessing suggested we were, and it is hardly possible to confute him with statistics. Only from our innermost experience can we decide whether we have been made more proof against

[1] *The Laocoon and Other Prose Writings of Lessing*, tr. and ed. W. B. Rönnfeldt (The Scott Library), London, n.d., p. 210.

Falstaff's lechery or Ford's jealousy or Page's eye to the main chance by seeing *The Merry Wives of Windsor*. This indeed is not comedy on the highest level, but it functions the more directly as an example. Our answer must surely be that what we gain from it, or any similar comedy, is an illusory sense of security, very similar to the sense of superiority that in David Storey's novel the audience got from seeing and listening to Gormless Gordon the gump from Gorseforth. It is, in fact, the smaller kind of comedy that leaves us feeling fortified and comfortably near to society's heart, that provides us, we may say, with a cheap and vicarious purge.

There is a more ambitious kind of comedy that refuses us the purge. It may be seen in *Measure for Measure*, where Vienna is as corrupt at the end as at the beginning, and we are driven to recognise that the Duke has failed in his original purpose of making the law more respected: merely to forgive everyone (or, rather, to use a forced marriage as the only punishment) is no likely way of making all the world follow the path of virtue; or in *Volpone*, where the punishment of eccentricity is so extreme as to bring the judgment of the court into question (even though Jonson was clearly unhappy about this and tried to defend its severity in his dedication of the play to the two Universities); or in *The Alchemist*, where Lovewit gets the young widow and all the booty, where Face puts his master in his debt, where Subtle and Doll lose only their expectations, and where the substantial losers are those of inferior cunning. In these plays we have no easy feelings that others are doing penance for our folly: we are brought up against the concept of a prevalent total injustice, we are perplexed rather than consoled or directed. And so it is too in the major comedies of more recent times. The characters of Chekhov's plays or of Ibsen's *The Wild Duck* are demonstrably given to folly, but they are not simply brought back into line with what society requires of them: like Malvolio in *Twelfth Night*, they persist, they maintain a challenge. Gregers Werle will go on being the thirteenth at table; the three sisters, abandoning the dream of getting to Moscow, will hope that some day beyond their time things will be differently ordered. In such comedies, and surely they are the most ambitious, there is no *catharsis*: at their endings we look about in uncertainty, we see that there have been no scapegoats to take away our guilt.

So, in the final analysis, we shall find that true of tragedy. We do indeed pause for contemplation and assimilation, we are quiet at the play's end. But it has already been seen how resentful we must be of a system of things that allows us to find vicarious atonement, that indeed demands of us any atonement. The hero dies, but does that help us for more than a moment of exhilarating admiration and contentment? It may be useful here to compare the 1607 and 1641 texts of Chapman's *Bussy d'Ambois*. In 1607 the play ended with the Friar's ghost speaking admiring words of the dead Bussy:

> Farewell brave relics of a complete man:
> Look up and see thy spirit made a star,
> Join flames with Hercules: and when thou set'st
> Thy radiant forehead in the firmament,
> Make the vast continent, crack'd with thy receipt,
> Spread to a world of fire: and th' aged sky,
> Cheer with new sparks of old humanity.
>
> (V. iii. 268–74)[1]

But in the version of 1641, as revised by the author himself or by another, these words are pushed back and are followed by a final colloquy between the cuckold Montsurry and his wife Tamyra: they are unable to come together in peace as Bussy urged them to do and they '*Exeunt severally*'. In this they symbolise our inability to find a lasting peace through Bussy's death, or Lear's or Macbeth's or the Duchess of Malfi's. It is as if the purge is offered and we, like Montsurry and Tamyra, have refused it.[2]

[1] *Bussy d'Ambois*, ed. Nicholas Brooke (The Revels Plays), London, 1964.

[2] I am not here suggesting that the 1641 revision is or is not in this instance Chapman's.

But, of course, we only ultimately refuse it. There is a moment of calm, there is a short-lived sense that the scapegoat has taken our cares away. Still, it would be the worst sort of humiliation if we allowed that to operate for long. We should be putting ourselves on the level of the comedian's audience in David Storey's *Radcliffe*. Indeed it seems legitimate to believe that the worst kind of response to tragedy is that where we feel most fully purged. We may be given a sense of human vulnerability, and because we in our littleness share this with the great figures of the stage we may be, in a facile way, consoled; or, alternatively, the details of what the stage presents may in retrospect seem to have some vague relation to fragments of our own experience and can thus encourage self-pity. Because the dramatised version of experience has been on a level of theatrical grandeur, we either can feel for the time being freed because we seem in comparison too little to be important, or can feel exalted because briefly identified with gods and heroes and can therefore carry away the notion of grandeur for a time. But the littleness is humiliating, the grandeur false, the self-pity enfeebling. Italo Svevo in his novel *Senilità*[1] has noted how these things operate as Amalia Brentani and her brother attend a performance of a Wagner opera:

> Her pain, absorbed into the music, took on a fresh colour and a greater significance, though at the same time it became simpler and purified of all that had defiled it. She was little and weak, and she had been beaten; how could she have hoped to have survived? She had never before been so resigned, so free from all anger. She felt she wanted to go on crying quietly and make no sound. Here in the theatre, of course, the solace of tears was denied her. But she had been wrong to say she did not understand the music. That magnificent stream of sound signified the whole of human destiny; she saw it pouring down an incline, its path shaped by the unequal conformation of the ground.

[1] Published 1898; the relevant passage is to be found in the translation by Beryl de Zoete, *As a Man Grows Older*, Harmondsworth, 1965, pp. 130–1.

Now it would flow in a single cascade, now it would be divided into a thousand smaller ones, all coloured by an ever-changing light, and by the reflections which objects cast upon it. There was the harmony of sound and colour which held the epic fate of Sieglinde, but also, insignificant though it was, her own, the end of a part of life, the withering of a single twig Her fate was no more to be wept over than that of the others; it deserved the same tears—no more; and the ridicule which had so cruelly oppressed her found no place in that picture which yet was so complete.

Her companion was familiar with the music, he knew how all those sounds were produced and how they were put together, but he did not succeed in getting so near to them as Amalia. He thought that his own passion and pain would immediately have clothed itself in the imagination of the composer. But no. For him, those who moved upon the scene were gods and heroes who transported him far away from the world of his sufferings. During the intervals he sought in vain among his memories for some experience which would have merited such a transformation: he could not find it. Had he perhaps found a cure in art?

When he left the theatre after the opera was over he was so full of this hope that he did not notice his sister was more cast down than usual. Filling his lungs with the cold night air, he said that the evening had done him a great deal of good. But as he went on chattering in his usual way about the strange calm which had pervaded him, a great sadness filled his heart. Art had only given him an interval of peace, and it would not be able to give it to him again, for now certain fragments of the music which had remained in his mind were already adapting themselves too perfectly to his own sensations, his self-pity, for example, and the sympathy which he felt for Angiolina or Amalia.

Thus to find relief through projection on to the tragic hero is a trifling and only fleetingly consolatory exercise We may say, as in effect Amalia in this novel, 'This only increases my sense of insignificance, for I cannot aspire to the grandeur of desolation that is here presented', or, like her brother, we may ultimately debase the level of tragedy to that of our small concerns. Attention, indeed, should be drawn to the fact that not all audience-responses to a major tragedy will be equally complex, equally mature.

And though it would have been inappropriate for Italo Svevo to have touched on it here, with the characters whose thoughts he was presenting, there is the further question: is any theatre representation free from the grandiose, the speaking-too-much? The greatest tragedy does indeed hint this doubt, and presents its heroes as by no means immune from our ridicule. So Othello takes too sanguine a view of himself even at the end, Lear boasts of an old man's strength in killing the slave whom he found hanging Cordelia, Macbeth gets a boyish sort of atonement in dying with a sword in his hand, Hamlet has a very human sensitivity about his reputation. Tragedy at its greatest denies a total sureness of greatness to the protagonist; by refusing us a scapegoat free from the pretence that he is the perfect victim, it will not let us forget our own frailty. The typical and greatest tragic hero is, as Aristotle put it,[1] greater than we are; he is also a man who talks too much.

So indeed it has appeared to Arthur Miller, a major playwright of our time: when he has turned from the play to the short story, he has felt that he has been able to show, 'amid the immodest heroics of the day', the things which 'for one reason or another do not belong on a stage'. He has gone on to refer to 'the theatrical tone of voice, which is always immodest, at bottom'. For the playwright 'is a per-former *manqué*'.[2]

If this is implicit in Shakespeare and his major con-temporaries, it is explicit in Marlowe. His Tamburlaine dies with an illusion of a continuing empire through his sons; his Faustus is damned without a hint that the world is the better for it; his England after Edward II and his Malta after Barabas and his France after the massacre and the death of the Guise are merely extended areas for striving. We may

[1] *The Poetics*, Chapter II.

[2] *Don't Need You Any More: Stories*, New York, reprinted 1968, pp. ix–x.

indeed wonder if Marlowe wrote tragedy in *Edward II*[1] and *The Jew of Malta* and *The Massacre at Paris*: tragedy does seem to demand, in however qualified a way, the sense of a major loss in an individual's ending, a sense of identification with him, a sense that his burden in some sense represents ours, even though on full consideration we may come to see that in his dying he has not diminished the weight we carry. In *Tamburlaine* and *Faustus* we do recognise tragedy, though not a tragedy that even momentarily consoles us. Part of the measure of Marlowe's independence is his frank refusal to let us feel purged: we pause for a moment to assimilate what has been put before us, but we know the hero's destruction is not a rite performed on our behalf: we cannot accept it even for an instant as vicarious atonement. Nor can we, after all, accept Hamlet's or Macbeth's or Othello's or Lear's or the Duchess of Malfi's. If we have any self-respect, we shall not for long be lulled by the music of the words or by the suggestion of a scapegoat rite. What tragedy finally has to offer us is the ability to see beyond purgation, to see that it is not possible, to see that we remain unpurged even if Lear dies. The smaller kind of comedy offers us reconciliation to society. Tragedy or the more ambitious kind of comedy subsumes and transcends *catharsis* and takes us to a level where we refuse to let our representative be humiliated without protest, without shame that we have for a moment acquiesced in his fall. There is, in the final analysis, nothing fortunate about it.

[1]This is implicit in my article '*Edward II*: Power and Suffering', *Critical Quarterly*, I (Autumn 1959), 181–96. But I have since come to think that in this instance, though not in *The Jew* or consistently in *The Massacre*, Marlowe does achieve a tragic effect through the very intensity of the suffering presented. We do not care whether a man is 'great' or not if his agony is acute in the way Edward's is—from the end of Act IV and of course supremely in his death-scene. Nevertheless, Marlowe's tragedy is here of an unusual kind.

Chapter 8

ELIZABETHAN AND
JACOBEAN

THERE was, of course, no revolution in 1603. An ageing queen died and was succeeded by a learned, energetic, self-important and sometimes foolish king from Scotland. In general policy the two sovereigns were not far apart: domestically they stood for compromise, an attempt to avoid or postpone open strife among the increasingly hostile factions into which English society was splitting up; both Elizabeth and James believed it was in the country's interest, and not merely a matter of divine right, for the central government to be strong; both tried to keep out of continental wars, Elizabeth fighting Spain only when she had to and James withholding English participation in the Thirty Years' War. Elizabeth did indeed in her own lifetime achieve the status of a myth and a wonder, so that it was possible for Spenser to celebrate her glory in Gloriana and her personal remoteness in Belphoebe, and for Chapman in *Bussy d'Ambois*, written shortly after her death, to make the French king and his courtiers pay tribute to her as 'the rarest Queen in Europe'. It was not so easy for James to be seen as the very archetype of monarchy. Men did look back to the queen's days with regret, and contrasted the disorder of James's court with the decorum of Elizabeth's. This is implied in the passage from *Bussy d'Ambois* already referred to (I. ii. 1–55); we may note also Sir John Harrington's remarks on the entertainment offered to James and Christian IV of Denmark at Theobalds in 1606: Harrington marvelled at 'these strange Pageantries' where both Christian and the performers were drunk and commented on the deterioration of court-manners since the queen's time.[1] Even so, in courtly entertainment and public pageant the new king was celebrated as a wonder, and on occasion a writer for the public theatre could make his contribution to the homage. *Henry VIII* was written by Shakespeare, or by Shakespeare and Fletcher, around 1613, when James had been on the throne for ten years, and we shall remember that it ends with Archbishop Cranmer speaking a prophecy over the infant Elizabeth at her christening. The future queen's greatness is celebrated, and then Cranmer proceeds to speak of the man who will succeed her:

> Nor shall this peace sleep with her; but as when
> The bird of wonder dies, the maiden phoenix,
> Her ashes new create another heir
> As great in admiration as herself,
> So shall she leave her blessedness to one –
> When heaven shall call her from this cloud of darkness –
> Who from the sacred ashes of her honour
> Shall star-like rise, as great in fame as she was,
> And so stand fix'd. Peace, plenty, love, truth, terror,
> That were the servants to this chosen infant,
> Shall then be his, and like a vine grow to him;
> Wherever the bright sun of heaven shall shine,
> His honour and the greatness of his name
> Shall be, and make new nations; he shall flourish,
> And like a mountain cedar reach his branches
> To all the plains about him; our children's children
> Shall see this and bless heaven.   (V. v. 40–56)

We may perhaps discern a touch of barely withheld dramatist's irony in Henry VIII's comment: 'Thou speakest wonders'; but we cannot deny that the author of the passage has worked strenuously to keep up the tradition. And even if one had to be a little strenuous in this, there may also have been a feeling of some genuine relief with the change of sovereign. The men of the queen's time had always

---

[1] John Nichols, *The Progresses, Processions, and Magnificent Festivities of King James the First*, 1828, II, 72–3.

worried about the succession, and the danger of a return to civil strife, or of the destruction of the Elizabethan establishment in religion, was fairly continuously present to them. Political insecurity was a theme for the dramatists from Sackville and Norton's *Gorboduc* in 1561 to Shakespeare's histories more than thirty years later. Now there seemed no reason to anticipate an overthrow of government. The Stuarts seemed firmly in, and if James was not easy for all men to like, they did enthusiastically esteem the young Prince Henry who, until his early death in 1612, was looked to as James's certain successor.

And there was, after all, some relief in not having a demi-goddess at hand. We can find an interesting change of tone in the prologues addressed to the sovereign at court if we compare the ways in which John Lyly and Ben Jonson wrote them. Lyly, introducing his *Campaspe* to the queen in 1584, compared her to the gods who 'supped once with poore Baucis': in watching the play, she is condescending as remarkably as Jove and Mercury had done. Jonson in 1614 welcomed the king to his *Bartholomew Fair*, telling him that he 'must expect' the men and the language appropriate to the play's subject, and offering for the king's 'sport' a sight of the Puritans—'your land's Faction'—'whereof the petulant ways Yourself have known, and have been vex'd with long'. This prologue ends by promising the king that his enjoyment of the play will constitute a 'fairing' — that is, the present that any boy or girl expected when taken to a fair. There is a kind of *camaraderie* here: you have had trouble with the Puritans, and so have we, the players and playwrights; you like fairs as we do; you must, in watching the comedy, direct your expectations in the right way. That this was not simply a personal difference between Lyly and Jonson can be seen from the tone of homage in the epilogue addressed to the queen that Jonson wrote for *Every Man Out of his Humour* (1599).[1] If James could be splenetic and rash,

[1] In Charles's time it was a different matter: Shirley in *The Triumph of Peace* (1634) is not untypical of Caroline masque-writers in the reverence of his tone.

if James's conduct ultimately contributed to the sharpening
of divisions in the land, if it was wished that he would not
make so many Scottish knights, there was nevertheless a
feeling of living in a more normal world than before. And
that encouraged the growth of satire and of the deeper kind
of questioning that takes on a tragic tone. The crisis of the
succession was over, the demi-goddess (to whom all praise,
men said) was gone: now attention could be turned more
freely to the sickness endemic in society and the paradox
of the human condition.

Yet it must be emphasised that this was a matter of
'encouraging the growth' of things already existent. There
was plenty of verse-satire while Elizabeth was on the throne,
and indeed the increase of satire in the playhouse at the
turn of the century has often been attributed to the ban on
the publication of satiric writings in 1599. Even the drama,
however, could exhibit a satiric strain substantially before
that. Robert Wilson the actor in *The Three Ladies of London*
and its sequel *The Three Lords and the Three Ladies of London*,
both written in the 1580s, could show London society
dominated by the figure of Lucre with her servants Usury,
Dissimulation, Fraud and Simony. And about 1591 Robert
Greene put into his generally romantic *James IV* a scene
where a Divine, a Merchant and a Lawyer accuse each other
of being responsible for the sick condition of the country.[1]
Corruption in law, faction in religion, and an instability
due to the rising and falling of families as gold changed
hands with, it appeared, ever-increasing rapidity each of
these in turn is seen in turn as the root of the ill, and the
three debaters are sharp in the blame they fix on one another.
This play also contains a servant, Andrew, who knows his
master to be a villain and who despises himself for contributing to roguery, for his own acceptance of humiliation. He
finds relief in satirical asides and ambiguous responses,
anticipating the Flamineo of *The White Devil*, the Bosola of
The Duchess of Malfi, the De Flores of *The Changeling*.

[1] See above, '*Catharsis* in English Renaissance Drama', p. 130.

And of course the Elizabethan years made a signal contribution to tragedy in the plays of Marlowe. No Jacobean went further than Marlowe in his exposure of the vanity and cruelty that go along with conquest and power; no one exceeded him in his intimate concern with physical suffering, most notably in the death-scene of *Edward II*;[1] no one else in the Elizabethan or Jacobean years so fully and frankly faced the idea of damnation for the central figure of a tragedy, taking the hypothetical situation of a human being in hell as part of the tragic writer's datum.

Even so, it is around the turn of the century that the drama's concern with the satiric and the tragic grows generally prominent. Jonson's *Every Man in his Humour* is acted in 1598 and is followed by the three plays he called his 'comicall satyres'; Marston, a new playwright, has his *Antonio and Mellida*, predominantly satiric, and its sequel *Antonio's Revenge*, predominantly tragic, acted in these years of Jonson's beginnings. Shakespeare contributes his *Troilus and Cressida* to the satiric kind of writing soon after. And, if tragedy almost sleeps for a while after Marlowe, it achieves sudden authority again in *Hamlet*, probably written, in the form in which we have it, in 1601. Then, with the new king's reign, there comes a remarkable leap forward in both of these dramatic kinds. In tragedy James's years saw *Othello, Lear, Macbeth, Antony and Cleopatra, Coriolanus, Timon of Athens; Bussy d'Ambois, Byron's Conspiracy and Tragedy, The Revenge of Bussy, Chabot; The Second Maiden's Tragedy; The Atheist's Tragedy, The Revenger's Tragedy; Sejanus, Catiline; The White Devil, The Duchess of Malfi; A Woman Killed with Kindness, The English Traveller; The Maid's Tragedy, Valentinian, Thierry and Theodoret; The Changeling, Women Beware Women*. These are not equally great tragedies, or equally good plays, but every one of them has a measure of authority, stage-worthiness, and intellectual honesty. They were all written within approximately twenty years, and collectively they constitute the reason for our speaking of 'Jacobean tragedy' as a

---

[1] See above, '*Catharsis* in English Renaissance, Drama', p. 147.

special variant of tragic drama, as a graspable entity. Tragedy, sporadic and various in Elizabeth's reign, moving hesitantly towards the supreme experiment of *Hamlet* almost at the end of that reign, becomes assured and in large measure homogeneous when her successor arrives.

Satiric comedy exists in its own right in Jonson and Shakespeare and Marston and Chapman and, with a difference, in Beaumont and Fletcher, but it also made a strong contribution to the total character of Jacobean tragedy. Whenever the tension is relaxed in the tragic writing of these years, it is likely that we shall find the dramatist turning his lash or his probe on the follies and vices of men in society, inviting our mockery of the ludicrous figures they present. Thus Kent in *King Lear* can delight in the series of insults that he feels Oswald has deserved:

> A knave, a rascal, an eater of broken meats; a base, proud, shallow, beggarly, three-suited, hundred-pound, filthy, worsted-stocking knave; a lily-liver'd, action-taking, whoreson, glass-gazing, superserviceable, finical rogue; one-trunk-inheriting slave; one that wouldst be a bawd in way of good service, and art nothing but the composition of a knave, beggar, coward, pander, and the son and heir of a mongrel bitch; one whom I will beat into clamorous whining, if thou deny'st the least syllable of thy additon. (II. ii. 15–26)

And when Cornwall asks for the reason for the brawl, we have this exchange:

> *Corn.* Why dost thou call him knave? What is his fault?
> *Kent.* His countenance likes me not.
> *Corn.* No more, per chance, does mine, nor his, nor hers.
> *Kent.* Sir, 'tis my occupation to be plain:
> I have seen better faces in my time
> Than stands on any shoulder that I see
> Before me at this instant. (II. ii. 95–101)

Then Cornwall, commenting on this, describes Kent as one of those malcontents who claim a privilege for simple vituperation on the grounds of plain-speaking. And here is Bosola in *The Duchess of Malfi* abusing the unnamed Old Lady

F

and the old courtier Castruchio, and beginning with a reference to the Old Lady's closet:

> One would suspect it for a shop of witchcraft, to find in it the fat of serpents, spawn of snakes, Jews' spittle, and their young children's ordure—and all these for the face: I would sooner eat a dead pigeon, taken from the soles of the feet of one sick of the plague, than kiss one of you fasting. Here are two of you, whose sin of your youth is the very patrimony of the physician, makes him renew his footcloth with the spring and change his high-prized courtezan with the fall of the leaf: I do wonder you do not loathe yourselves. (II. i. 35–44)[1]

I have said this happens when the tension is relaxed, but there is no suggestion of something extraneous coming into the play here. We have only to think of Lear's words on hypocrisy and social corruption when, as the disguised Edgar puts it, 'reason in madness' is what he utters, or of the Duchess of Malfi's mockery of rich men and of her grotesque association of luxury with death, as she waits for the moment when she herself is to be brutally strangled, or of Vindice's savage speech on promiscuous lust as he views the painted lips of his dead love's skull: in each instance the satire is at the heart of the tragic idea. The corruption and the folly are held in the mind simultaneously with the feeling that mankind is somehow dignified by the appalling condition into which it has been born.

A group of dramatic tragedies that are in large measure akin to one another and that incorporate satire is principally what has made James I's reign endure in human memory—along with the translation of the Bible associated with his name. The contrast between the two achievements is characteristic of the time. And Shakespeare made his contribution to Jacobean tragedy, as he had notably helped in the more varied writings of the Elizabethan years, giving the theatre his histories, romantic or neo-Plautan comedies, and brilliant though uncertain essays in tragedy. Thus, if

[1] *The Duchess of Malfi*, ed. J. R. Brown (The Revels Plays), London, 1964.

we look for a mode of distinguishing 'Elizabethan' from 'Jacobean', and concentrate our attention on the drama, we shall find it first of all in the rise to greater authority in tragedy and dramatic satire, and secondly in the degree of homogeneity in the best plays of the Jacobean years as a total group, whatever their immediate Kind may be. There is no impassable gulf between *Volpone* and *The White Devil* and *The Winter's Tale*, and we shall realise that more clearly if we think of the gulfs between *Doctor Faustus* and *Love's Labour's Lost* and *Woodstock*,[1] or the lesser gulfs a few years later between *Julius Caesar* and *The Shoemakers' Holiday* and *Antonio and Mellida*.

But homogeneity is a Jacobean characteristic in another and more subtle way. To illustrate this, we may take two speeches already referred to in plays by contemporaries of Shakespeare: the Duchess of Malfi's speech before her murder, and Vindice's speech as he surveys his dead love's skull just before he is to use it to lead her poisoner the Duke into lust and thence to death. These speeches have been noted as bringing the tragic and the satiric into close relation: now I want to suggest that each gives us in its brief compass an image of its play's total world. For the Duchess it is necessary to quote the interspersed comments from Bosola and the Executioners, but her words virtually constitute a single speech:

Bos. Yet, methinks,
The manner of your death should much afflict you,
This cord should terrify you?
   Duch.   Not a whit:
What would it pleasure me to have my throat cut
With diamonds? or to be smothered
With cassia? or to be shot to death with pearls?
I know death hath ten thousand several doors

---

[1] Earlier in the popular theatre the distinction was less sharp: the moral tragedy (e.g., *Cambises*) and the play of romance (e.g., *Sir Clyomon and Clamydes*) could be close together. Then greater self-consciousness led to the temporary separation of Kinds noted here.

For men to take their exits; and 'tis found
They go on such strange geometrical hinges,
You may open them both ways:—any way, for heaven-sake,
So I were out of your whispering:—tell my brothers
That I perceive death, now I am well awake,
Best gift is they can give, or I can take.
I would fain put off my last woman's fault,
I'd not be tedious to you.
    Execut.   We are ready.
    Duch. Dispose my breath how please you, but my body
Bestow upon my women, will you?
    Execut.   Yes.
    Duch. Pull, and pull strongly, for your able strength
Must pull down heaven upon me:
Yet stay; heaven-gates are not so highly arch'd
As princes' palaces, they that enter there
Must go upon their knees. —[*Kneels.*] Come violent death,
Serve for mandragora to make me sleep!
Go tell my brothers, when I am laid out,
They then may feed in quiet.   (IV. ii. 213–37)

She is not terrified by the cord which will strangle her; she makes fun of the thought of luxury, thinking how ironic it would be to use diamonds or pearls or cassia in the act of death; she sees the universe as a prison with ten thousand swinging doors, doors that can be pushed from man's side as well as God's; she longs to escape from the 'whispering' that her brothers' intrigues and the deeds of their servants have made a perpetual accompaniment to her last days; she jests at her own talkativeness, as a 'woman's fault'; she wants her dead body treated with respect; she thinks of the impertinence of the executioners, who will dare to force heaven to accept a newcomer; she glances satirically at the pomp of princes' palaces; she sees a paradox in the violence of the killing and the stillness it will induce, as if the wrenched and racked body had been given a sleeping-draught; she ends with an image of her sated brothers feeding in quiet, like beasts of prey after a kill. There is dignity and wit, a measured fear and resolution; there is a strong sense of the actual, the concrete (the diamonds, the swinging

doors, the princes' palaces, the wild beasts); there is a
guarded act of faith, a reverence for the body, a mockery
of her own small failing, that of talking too much; there is
final reproach for her brothers rather than an assured
turning to the heaven she kneels to.

And now we may see Vindice in *The Revenger's Tragedy*
contemplating the image of tricked-up death:

And now methinks I could e'en chide myself
For doting on her beauty, though her death
Shall be reveng'd after no common action.
Does the silk-worm expend her yellow labours
For thee? for thee does she undo herself?
Are lordships sold to maintain ladyships
For the poor benefit of a bewitching minute?
Why does yon fellow falsify high-ways,
And put his life between the judge's lips,
To refine such a thing? keeps horse and men
To beat their valours for her?
Surely we are all mad people, and they
Whom we think are, are not; we mistake those:
'Tis we are mad in sense, they but in clothes.
   Hipp.   Faith, and in clothes too we, give us our due.
   Vind.   Does every proud and self-affecting dame
Camphor her face for this? and grieve her maker
In sinful baths of milk, when many an infant starves
For her superfluous outside—all for this?
Who now bids twenty pound a night, prepares
Music, perfumes and sweetmeats? All are hush'd;
Thou mayst lie chaste now. It were fine, methinks,
To have thee seen at revels, forgetful feasts,
And unclean brothels; sure 'twould fright the sinner,
And make him a good coward, put a reveller
Out of his antic amble,
And cloy an epicure with empty dishes.
Here might a scornful and ambitious woman
Look through and through herself;—see, ladies, with false forms
You deceive men, but cannot deceive worms.   (III. v. 69–98)[1]

---

[1] *The Revenger's Tragedy*, ed. R. A. Foakes (The Revels Plays), London,
1966.

He begins by glorying in his brilliant plan of revenge, though he has come to feel that the cause of the revenge is no longer sufficient. Why do men do such notable and dangerous things for the sake of a woman—losing their patrimony, taking to highway robbery, making a great show, just as the silkworm lives only to adorn a woman's body? We are madder than mad people except in our mode of dressing, and perhaps mad in that too. Then he turns to the woman's activities: why does she labour at the toilet and let the poor starve, when in the end she must become what his own mistress now is? Would it not be a good thing to have a skeleton on show at revels and in brothels? What Tourneur offers is not only satire: few writers can more vividly present, more sympathetically enter into, the life of sensuality and danger and wastefulness that he overtly mocks. Vindice mocks also at himself, for he has loved and now he is putting 'his life between the judge's lips' because this woman has died and he wants revenge; at least the revenge will be no common one; all life is deception, until death reveals the truth by taking away the beauty of the flesh and then the flesh itself. All the irony of the play is here, all the disgust, and all the relish in spite of the disgust. Even the chastity of Castiza, Vindice's sister, is represented, for the woman whose skull he looks on was chaste too and died for it. Even a basic dubiety about virtue is hinted in the paradox 'good coward'.

Of course, not every passage in *The Duchess of Malfi* or *The Revenger's Tragedy* or in any other play of this time can act in this manner of bringing the play's total world before us. However, many dramatic speeches of these years do work in this way, and, even where the scope of a speech is altogether slighter, it is likely to have coherence with the play's total manner and total world. The point can be best made by a contrast or two from the Elizabethan years. All the world knows Portia's lines on the quality of mercy: they are eloquent, and they introduce an idea that is relevant to and important in the play's thought-structure. But they exist on an abstract level, with their talk of sceptres and a rain that is conventionally 'gentle'. There is no sense of the human involvement and physical contact that are evident in other parts of the play, or of the impetus to revenge that humiliation and envy can give, no recognition of the non-homogeneous society that the play illustrates in the presence of the Jews in Venice, no sense either of that partiality and arrogance that Portia, as judge, is not free from. Or we can think of Henry V's speech before Agincourt, with its impulsion to the winning of honour by taking on a superior enemy, the winning of a happy old age spent, at least on annual occasions, in showing one's scars and fighting one's successful battles over again. Once more there is eloquence divorced from a sharp consciousness of physical fact and divorced also from many important elements in the play's thought: the satire of ecclesiastics whose motives in giving a moral judgment are rendered suspect; the subdued complainings and questionings of the private soldiers John Bates and Michael Williams and the almost complete silence of their fellow, Alexander Court, as they wait for the morning of Agincourt to come; the thrice-repeated command to kill the French prisoners; the hanging of Bardolph; the moral shabbiness of some of the soldiers on Henry's side. I want to take up later the closer sense of the physical, the concrete, in Jacobean writing. What I am concerned with here is that the Elizabethan plays here mentioned, though indeed complex plays, exhibit their complexity through the juxtaposition of scenes and speeches rather than through the interpenetration of words.[1] Bertolt Brecht's relation to Shakespeare is often stressed to-day: that is at times an illuminating comparison, but it would be better sometimes to note that he is akin to the Elizabethans rather than the Jacobeans. Where he approaches Shakespeare's manner, it is the Shakespeare of the histories, the early and middle comedies, the early essays in the tragic. The Elizabethan way of writing is to put things together, the Jacobean way is to fuse. In Coleridgean terms, we pass from a drama of

---

[1] See above, 'A School of Criticism', p. 39.

the fancy to a drama of the imagination. Here of course Coleridge's distinction is used in a way he does not use it: his concern does not seem to have been total structure—at least in relation to a writing of substantial length—but rather the coming of 'poetry', the result of imaginative activity, from time to time into a structure appropriate to it.[1] Nevertheless, his terms may clarify the distinction that is here being made. *The Merchant of Venice* and *Henry V* owe their complexity to Shakespeare's exposing in turn the many facets of his polyhedron: we get a sense of the whole through the operation of memory. But throughout *The Duchess of Malfi* or *The Revenger's Tragedy* we are fully or marginally aware of the play's world, and, in the moments from those plays that I have specifically referred to, that world is—in its full extent and in vital physical terms— stamped on our minds.

To illustrate from Shakespeare this fusion, this imaging of the total play in a single speech, we may take Ulysses' words on emulation in *Troilus and Cressida*. Certainly this play was written before James was on the throne, but it has been noted that his accession only strengthened tendencies already in being: *Troilus and Cressida* is in manner and spirit far more Jacobean than Elizabethan. In commenting on Ulysses' long speech, it will be best to take it in sections of a few lines each:

> Time hath, my lord, a wallet at his back,
> Wherein he puts alms for oblivion,
> A great-siz'd monster of ingratitudes.
> Those scraps are good deeds past, which are devour'd
> As fast as they are made, forgot as soon
> As done.   (III. iii. 145–50)

The 'Time' that is mentioned is important throughout the play, bringing an end to Troy, an end to Cressida's love, and syphilis to Pandarus. Here it is presented as a begging pilgrim with a wallet for the 'alms' he gathers; but he takes

---

[1] See below, 'The Dramatist's Experience', pp. 222–3.

the alms, the 'scraps', to feed 'Oblivion', a huge and devouring monster that he, Time, serves. On the one side the imagery is sharply, sometimes grossly, physical ('wallet', 'back', 'scraps', 'devour'd'); on the other there are the faint and generalised 'good deeds', frail here in the extreme. Then Ulysses gives advice:

> Perseverance, dear my lord,
> Keeps honour bright. To have done is to hang
> Quite out of fashion, like a rusty mail
> In monumental mock'ry. Take the instant way;
> For honour travels in a strait so narrow
> Where one but goes abreast. Keep then the path,
> For emulation hath a thousand sons
> That one by one pursue; if you give way,
> Or hedge aside from the direct forthright,
> Like to an ent'red tide they all rush by
> And leave you hindmost;
> Or, like a gallant horse fall'n in first rank,
> Lie there for pavement to the abject rear,
> O'er-run and trampled on.   (III. iii. 150–63)

The sequence here is notable. 'Perseverance' sounds well enough, and is vivified by the image of keeping honour 'bright', so that we are made to witness a long and successful process of polishing: there is a touch of futility, but no more. Then the contrasted picture of the 'rusty mail', with the colloquial 'Quite out of fashion' and the antithesis of 'monumental mock'ry' bringing the *Denkmal*, the monumental itself, within the reach of the mockery. Better, surely, to spend one's time polishing than to hang on the wall in this fashion. Then a series of images which grow increasingly threatening: first, the narrow strait where only one man can pass at a time (the point being made through the paradox that only one may go abreast—an utterance as defiant of common usage as the alleged line in *Julius Caesar* that Jonson rebuked Shakespeare for[1]); then the thousand-strong

---

[1] *Discoveries* in *Ben Jonson*, ed. C. H. Herford and P. and E. Simpson, Oxford, VIII (1947), 584.

progeny of Emulation taking every opportunity to get past each other, the image of the 'ent'red tide' giving to their movement the dark authority of a force of nature; and then, with sharper concreteness, the horse 'O'er-run and trampled on'. Quickly we have moved from the decorous image of the armour-polishing to the disordered picture of destruction and pain: the transition is appropriate in a play whose action extends from the pageant of Hector's challenge to his murder when unarmed. Ulysses then relaxes for a moment into direct exhortation, reinforced by the slighting mention of a host's conduct with his guests:

> Then what they do in present,
> Though less than yours in past, must o'ertop yours;
> For Time is like a fashionable host,
> That slightly shakes his parting guest by th' hand;
> And with his arms out-stretch'd, as he would fly,
> Grasps in the comer. The welcome ever smiles,
> And farewell goes out sighing.   (III. iii. 163–9)

The host is comic, of course: he is 'fashionable', he stretches out his arms 'as he would fly' (we have all known such over-enthusiastic hosts), and the 'grasp' of his welcome is a parody of what a handclasp should indicate. The parting guest is a little comic, too, for his 'sighing' suggests he has accepted as genuine the greeting he had when he came. Moreover, the smile of 'welcome' is the host's, but the 'sighing' is the guest's: it seems that only at the moment of rejection is one allowed a separate existence. Yet the dominant impression left us is that both guests are depersonalised, reduced to the abstractions 'welcome' and 'farewell', made to seem mere figments of the host's extravagant creation—puppets, in fact, of Time. This flashy lord has become more destructive than the monster-attended pilgrim of the speech's beginning. In the lines that follow, Ulysses reaches the climax of his argument:

> O, let not virtue seek
> Remuneration for the thing it was;

For beauty, wit,
High birth, vigour of bone, desert in service,
Love, friendship, charity, are subjects all
To envious and calumniating Time.
One touch of nature makes the whole world kin—
That all with one consent praise new-born gawds,
Though they are made and moulded of things past,
And give to dust that is a little gilt
More laud than gilt o'er-dusted.
The present eye praises the present object.

(III. iii. 169–80)

The catalogue of abstracts is not arranged in chance fashion. Ulysses first names those things that we all, at our most sincere, value: beauty and wit, fairness of form and mobility of mind; and then the things that the world respects because they exercise sway in society: high birth, for the authority it gives, vigour of bone, for its immediate power and threat, desert in service, for the claim it implies; and then those things that, not necessarily at our most sincere but at our most thoughtful, we put highest: love and friendship and charity, the particularised and the generalised forms under which we recognise the fact of humankind. All these things, he says, are destroyed by the Time that has been seen as a begging pilgrim and a fashionable host; but now he is a master, a master who envies and slanders his servants, a master whose conduct belies his own mastership. The line that follows, so frequently quoted out of context and misapplied—'One touch of nature makes the whole world kin'—is especially important for this play which insists so often on the bonds that link Trojans and Greeks, Helen and Cressida, Ajax and Achilles, Hector and Paris, Pandarus and Troilus, Thersites and Ulysses, Agamemnon and Calchas. Here the line relates to a general lust for novelty and trumpery, for the eye-catching and the modish, and then Ulysses remarks on man's disregard of memory, the eye having room only for what it sees now. In the last words of the speech, its lessons are applied to its hearer, Achilles,

163

for Ulysses' whole object has been to make Achilles return
to the war:

> Then marvel not, thou great and complete man,
> That all the Greeks begin to worship Ajax,
> Since things in motion sooner catch the eye
> Than what stirs not. The cry went once on thee,
> And still it might, and yet it may again,
> If thou wouldst not entomb thyself alive
> And case thy reputation in thy tent,
> Whose glorious deeds but in these fields of late
> Made emulous missions 'mongst the gods themselves,
> And drave great Mars to faction.   (III. iii. 181–90)

There is flattery here ('thou great and complete man'),
there is lying ('all the Greeks begin to worship Ajax'), there
is the conditional half-promise of a return to reputation,
and finally there is praise of the emulation that has recently
involved gods as well as men on the field of battle. The
antitheses between the eloquence of the speaking and its
petty aim, between the frightening general vision and the
speaker's cunning, are in accord with this play in which the
facts of love and death and solitude are shown dressed up in
a parody of the chivalric code.

It can be taken, of course, as the 'degree' speech in reverse.
There Ulysses had presented emulation as the destroyer,
the bringer of anarchy; here, to win his point with his present
hearer, he praises that same emulation and sees it as necessary for a man who will avoid being trampled on or, at the
least, hung like a rusty coat of mail on a wall. Yet the notion
of degree is implied in the second speech as it is explicated
at length in the first. If nature is the chaos of the speech we
have been looking at, if Achilles' emulation has been stretch-
ed even to the level of the gods, if Time of its being is all-
destroying and apparently malicious, then the only tolerable
way of living is to go against nature, to impose an order
upon it (however imperfect and fugitive that order may
be), to practise the love, friendship, charity that cannot last

but are recognised as good while they exist.[1] I have argued elsewhere that the tone and the very length of the 'degree' speech indicate that the notion of a cosmic, a divinely appointed order had become, by the time this play was written, a matter for defence and debate.[2] The ultimate implication of that speech is surely: if we have not been given a harmonious universe (and perhaps we have not), we must make what we have as harmonious as we can. This notion strongly underlies the 'emulation' speech too, as it underlies the whole play—a play in which good things are vulnerable but are not the less good for that. The human kindness of Hector (though his thoughtless playing at war may mask it), the actual, though frail and occasionally a little tawdry, love of Troilus and Cressida, the intellectual shrewdness of Ulysses (however much it is tainted by his political use of it): these things can provide the bases for an acceptable mode for living, even if those bases have continuously to be made anew. There are, in fact, affirmations in this play's totality, though they are not sanguinely offered. And they are strongly implied in the catalogue that Ulysses gives of the things that are 'subjects all To envious and calumniating Time'. A horse does not have to be trampled on every time we see a fellow man on its back, a host may choose to be unfashionable for a while, we may from time to time remember the past (to value, not to misuse it in the search for false analogues) and refuse the eye-catching properties of the present. It is true that we shall not manage these things often, but the satire of Ulysses' speech goes along with a recognition of the things that emerge from, and survive, the satire elsewhere in the play.

I now want to consider a matter so far referred to only in

[1] One may compare the words of a character in Douglas LePan's *The Deserter* (1964): 'I don't believe there's any order to be found in the world, or in the ridiculous small world of the state. Still I think it's worthwhile to try to give it one, and an order as free and generous as possible.' (p. 56)

[2] *Shakespeare's Tragedies and Other Studies in Seventeenth Century Drama*, London, 1951, p. 31.

passing, the much sharper sense of the concrete, the lived-with, in the Jacobean style. Portia confronting Shylock, Henry V at Agincourt, as we have seen, used a language that depended in large measure on generalised conceptions— 'gentle rain', 'show his scars'—which do not make us feel the rain or see (or feel) the scars. I want to emphasise that I see nothing wrong with dramatic language of this kind. It is the language of French classical tragedy, with its talk of 'feu' and 'gloire': the famous 'Vénus tout entière à sa proie attachée' is unusual in bringing the goddess to this side of abstraction. The Elizabethans used much more familiar language than Racine did, but not even Marlowe makes us feel the edges and savours of things as a dozen Jacobeans can. If, for example, we take Tamburlaine's speech about the poet's inability to give total expression to the idea of beauty, we shall at once see that we are moving in a world of inert abstractions:

> What is beauty, saith my sufferings, then?
> If all the pens that ever poets held
> Had fed the feeling of their masters' thoughts,
> And every sweetness that inspir'd their hearts,
> Their minds and muses on admired themes;
> If all the heavenly quintessence they still
> From their immortal flowers of poesy,
> Wherein as in a mirror we perceive
> The highest reaches of a human wit—
> If these had made one poem's period,
> And all combin'd in beauty's worthiness,
> Yet should there hover in their restless heads
> One thought, one grace, one wonder, at the least,
> Which into words no virtue can digest.
> (Part I, V. ii. 97–110)[1]

First of all, the subject is generalised ('Beauty'), though that might not prevent the dramatist from making us feel its effect. Then we have a reference to the pens held by the

[1] *Tamburlaine the Great*, ed. U. M. Ellis-Fermor, London, 1930, revised 1951.

poets, but we do not get a sense of what it is like to be a poet holding a pen. The image of distillation ('the heavenly quintessence they still') is kept away from the concrete by the conventional 'immortal flowers of poesy' in the next line, and then the image of a mirror is only momentarily and casually taken up. Moreover, the 'hover' and 'digest' of the last three lines do not fit easily together and are separated by the inert catalogue 'One thought, one grace, one wonder'. It is true that this is not Marlowe at his best (and may indeed be deliberately magniloquent at Tamburlaine's expense), and that he gave us the idea of human beauty more vividly in Faustus' address to the spirit that had the appearance of Helen. But there, too, the only idea that makes a physical impact is the kiss. Troy and the Greek ships, the wearing of Helen's colours on the Dardan plain or before Wittenberg, the wounding of Achilles in the heel: these have the nature of embroidery, of eloquence. They are magnificent, of course, and the picture of 'the evening's air Clad in the beauty of a thousand stars' is splendid to look on. But what Marlowe does with both Tamburlaine and Faustus is to make them recite hymns in honour of beauty which are in strict counterpoint with what they say and do elsewhere in the play. If he had made the images as concrete as a Jacobean might have made them, both he and his characters and we would have become fully involved with the things and activities the separate speeches are concerned with. We should then be in no position to realise the plays' special kind of multi-significance, which depends on a uniform distancing of the physical world. The complexity of Marlowe's dramatic writing is not to be denied, but his eloquent language is simply one of the elements in a total, highly dramatic compound: it does not in itself embody the compound; it is subdued to its separate purpose.[1] For the

[1] T. B. Tomlinson, *A Study of Elizabethan and Jacobean Tragedy*, Cambridge, 1964, pp. 48–58, has observed the frequent non-concreteness and the non-interpenetration of Marlowe's language (which he finds 'sterile'): what he misses is that Marlowe is writing a different sort of drama from the Jacobeans.

Jacobeans, multi-significance showed itself at every moment, but if one writes plays which depend on juxtaposition rather than fusion one must hold the physical world at arm's length. 'One might almost say, her body thought', says Donne in *The Second Anniversary*, and this is indeed the Jacobeans' way. The full recognition of the senses is part of what the universe means to them: it operates throughout their writing, and makes possible their mirroring of the play's total world in the single utterance. For them, the total world is a matter of sensual response as well as intellectual apprehension. I have referred to the catalogue in Ulysses' speech on emulation:

> beauty, wit,
> High birth, vigour of bone, desert in service,
> Love, friendship, charity . . .

We may put it with Donne's catalogue in his sonnet 'At the round earth's imagin'd corners':

> All whom the flood did, and fire shall o'erthrow,
> All whom warre, dearth, age, agues, tyrannies,
> Despair, law, chance, hath slaine . . .[1]

He begins with two unique and large-scale events: of the remote past, the flood; of the remote future, the ending of the world by fire. Within that range of time, he lists the recurrent causes of death: war, then a result of war (dearth), then the dearth that regularly comes by nature (age), then, echoing the sound, one of the diseases age is subject to (agues), then another cause of trembling (tyrannies) which reaches back to make us see war and dearth and age and agues as tyrants too, and which is itself vivified by its proximity to the strongly physical 'agues'; then the despair that tyrannies can induce, then—following the suicide of despair—the murders of the law, then accidental death (chance) which reflects back on the law's arbitrariness. As Longinus put it, 'the words buttress one another'.[2] We can contrast Marlowe's

---

[1] John Donne, *Complete Poetry and Selected Prose*, ed. John Hayward, London, 1929, reprinted 1939.

[2] *On the Sublime*, Chapter XL. See above. 'A School of Criticism', p. 39.

catalogue at the end of the Second Sestiad of *Hero and Leander*, where Night, 'o'ercome with anguish, shame, and rage, Dang'd down to hell her loathsome carriage'. No increased value is given to 'anguish', 'shame' or 'rage' by their association here. This is addition, not fusion. Donne and Shakespeare, however, give us a kind of catalogue which demands that we respond to the exact order of the words and feel the way each plays on the others. Through this interplay of concept with concept and image with image, the distinctions and associations become a matter of sensual-cum-intellectual apprehension. The Elizabethans are grandly eloquent; the Jacobeans are more concerned all the time— hence the homogeneous patterning of their plays—with the sense of physical impact.

This is not uniformly true (it is, for example, less true of Chapman than of the other writers of comparable stature) and it did not last. Eliot's 'dissociation of sensibility' not only existed before the Jacobeans were writing but returned quickly with some of the Caroline writers. A new turn was given to drama in Charles's reign. Then the major playwrights were Massinger and Ford, and both of them are inclined to the language of abstraction and the dramatic structure that depends on juxtaposition. Not totally, of course, and it must again be remembered that all that is said here is said in general terms. And I am not denigrating Massinger and Ford in saying that their drama has not the characteristic Jacobean fusion and physical impact. Coleridge, writing on *Hamlet*, differentiated between the purely 'dramatic' passages where character in action is fully presented and those passages where the lyric or the epic mode is in evidence.[1] So Wolfgang Clemen sees Shakespeare's imagery as truly 'dramatic' only when it is not decorative but enters wholly into the play's total concept.[2] Derek

---

[1] *Coleridge's Shakespearean Criticism*, ed T. M. Raysor (Everyman's Library), 1960, I, 21, 25.

[2] *The Development of Shakespeare's Imagery*, London, 1951, p. 101.

Traversi[1] and Una Ellis-Fermor[2] have come near to demanding for true drama an Ibsen-like sense of causality underlying its structure. There was a time when one had to try to persuade students that the picture-frame stage of the late nineteenth century was not necessarily the best of all possible stages. Now, though probably not for long, one has to plead for a recognition of the kind of drama that does not depend on causality, that does not aim at fusion, that may use rhetoric rather than the attempt to bring into every key-passage the total feeling and imagined world of the play. We should be helped in this regard by the current popularity of Brecht, for he is after all much nearer to Marlowe than to Shakespeare.

Shakespeare, by virtue of being a major poet and dramatist, was a man standing apart from his age. No writer of distinction merely echoes what is current in his place and time. Indeed, no merely intelligent man does that: do any, even, of us accept wholeheartedly the common notions that our small world offers us? We must not think of a Shakespeare content with holding up an Elizabethan or a Jacobean 'world-picture'. We cannot be wholly free from current opinions, even in rebelling against them, but as individuals we shall make our variants upon them. Shakespeare inevitably did much more. He was not merely an Elizabethan or a Jacobean. Nevertheless, it can be useful to observe his degree of kinship with his fellows both before and after the year when Elizabeth died and James succeeded her. Until he wrote *Hamlet* and *Troilus and Cressida*, Shakespeare is predominantly an Elizabethan, dependent on the idea of juxtaposition and expressing himself primarily in terms of abstraction, using the conventional word and the mode of eloquence; but from *Hamlet* on he moves to close grips with the physical world and works towards a condition of fusion where the whole impact of the play is implied in every moment of it. He is the world's greatest dramatist, and we

---

[1] *An Approach to Shakespeare*, London, revised edition, 1956, p. 128.

[2] *Shakespeare the Dramatist and Other Papers*, London, 1961, pp. 47, 115–16, 120n.

are inclined to think that every dramatist should aim at doing what Shakespeare did in his maturity. But for a moment let us assume that his career ended with *Henry V* and *Julius Caesar* and *Twelfth Night*, around the year 1600. That would still give him seven more years of life than Marlowe had. It would still make him, doubtless, our greatest dramatist in English, though not the greatest dramatist in the world. Then our normal expectations would be in terms of dramatic juxtaposition rather than fusion, and we should not be so narrowminded in our criticism as we often are. The Elizabethans—particularly Marlowe and the Shakespeare of the years before 1600—were magnificent. It was hardly their fault that the Jacobeans, using a different but not necessarily a superior mode, were in some instances better. The difference in mode is something that makes itself more fully apparent if we study Shakespeare's contemporaries along with him. Some of those contemporaries were themselves major dramatists, but none of them straddles the gap between Elizabethan and Jacobean as manifestly as Shakespeare does. Because each of them belongs on one side or the other, they clarify the pattern of Shakespeare's own development.

Chapter 9

CONGREVE AND THE
CENTURY'S END

W HEN Dryden returned to the theatre after the Revolution, it was with a sense that the glory had departed.
The comedies of Etherege and Wycherley had all been written
before 1680; the careers of Otway and Lee were done; the
only notable dramatist to begin work in the 1680s was
Southerne, and he had yet shown no signs of unusual worth.
So, when *Don Sebastian, King of Portugal* was published in
1690, Dryden wrote in a melancholy strain on the exhausted
condition of the Restoration theatre and drama:

> Having been longer acquainted with the Stage, than any Poet
> now living, and having observed how difficult it was to please;
> that the Humours of Comedy were almost spent, that Love and
> Honour (the mistaken Topicks of Tragedy) were quite worn out,
> that the Theatres could not support their Charges, that the
> Audience forsook them, that young Men without Learning set
> up for Judges, and that they talk'd loudest, who understood the
> least: All these Discouragements had not only wean'd me from
> the Stage, but had also given me a Loathing of it. But enough of
> this: the Difficulties continue; they encrease, and I am still con-
> demn'd to dig in those exhausted Mines.[1]

What variations were possible on the Restoration comic
formula, what could bring to serious drama something more
than the setting-forth of stereotyped attitudes? Gone was
the excitement of Puritan-baiting under the protection of
Charles's court, gone the feeling of release and adventure
that came with the returning monarchy, gone even the dream
of magnanimity that informed the heroic play. To Dryden

---

[1] From the Preface. The quotation is from *The Dramatic Works of
John Dryden, Esq.*, 6 vols. (London, 1762).

after the Revolution the theatre was a source of livelihood, and he could yet work at it with skill, but in 1690 he could see no contemporary dramatist worthy of much respect and no sign of a new development that would bring the spirit of William's reign into dramatic focus. This state of things needs to be remembered when we look at the commendatory verses published with Congreve's *The Old Bachelor* (1693) and *The Double Dealer* (1694). Southerne and Bevil Higgons, introducing the earlier play, hailed the writer as Dryden's successor, and in well-known lines Southerne drew attention to the prevailing dearth of talent. Of Dryden's contemporaries he writes:

> His eldest *Wicherly*, in wise Retreat,
> Thought it not worth his Quiet to be Great.
> Loose, wandring, *Etherege*, in wild Pleasures tost,
> And foreign Int'rests, to his Hopes long lost:
> Poor *Lee* and *Otway* dead!

Dryden himself introduced *The Double Dealer*, claiming that the new playwright united the best qualities of Restoration comic writing:

> *In Him all Beauties of this Age we see;*
> Etherege *his Courtship*, Southern's *Purity;*
> *The Satire, Wit, and Strength of Manly* Wicherly.

In dramatic genius, Dryden averred, Congreve was Shakespeare's peer, and the tribute ends with a formal bequeathing of Dryden's own laurels.

We are not likely to go so far as Dryden in all he says here, but he may give us a clue to the understanding of Congreve when he says:

> *In Him all Beauties of this Age we see.*

Congreve brings the diverse and often conflicting elements of Restoration comedy into a unity. Even in *The Old Bachelor* we find nothing of that disharmony that marked Etherege's beginning in *The Comical Revenge* or Wycherley's in *Love in a Wood*. In regarding the earlier writers, moreover, it is

173

possible, though often difficult, to divide up their work into such classes as the 'comedy of manners,' the 'comedy of humours,' the 'comedy of intrigue':[1] this has no validity at all for Congreve. He tells us in his letter to John Dennis, 'Concerning Humour in Comedy,' that good comic writing depends on the exposure of individual eccentricity and that this needs to be combined with the exposure of fashionable affectations; he uses plot-devices from the intrigue-comedies, and this is indeed suggested in the prologue to *The Way of the World* when he makes Betterton claim on the poet's behalf: '*Some Plot we think he has.*' His characters belong for the most part to the stock-types of the age—men and women of wit and fashion; harmless eccentrics like Foresight and Heartwell; men and women amorously inclined despite their years, like Sir Sampson Legend and Lady Wishfort; unpolished intruders into London society, like Ben and Sir Wilfull Witwoud; women of light virtue; fops and would-be wits—but he so contrives his plays that the characters are not isolated targets but are seen in relation to one another and to their society as a whole. His manner is generally light and courtly, as Etherege's was at its best, but there is a deliberateness in his utterance that could remind Dryden of Wycherley. He makes great play with the passion for similitudes which the wits indulged; his own fineness of phrasing depends largely on the current antithetic sentence-structure and on repetitions of word or phrase in the Wycherley mode; but his writing is flexible, warm, and rich in copious, startling, and strangely happy flights of fancy. In *The Way of the World* we find almost every character speaking in character and yet sharing in the fineness and amplitude of Congreve's language. Here is Mirabel, when Millamant has left him with the behest 'think of me':

I have something more—Gone—Think of you! To think of a Whirlwind, tho' 'twere in a Whirlwind, were a Case of more steady Contemplation; a very Tranquility of Mind and Mansion.

[1] Cf. Allardyce Nicoll. *A History of English Drama 1660–1900*, I(Cambridge, 1952), 194–5.

A Fellow that lives in a Windmill, has not a more whimsical Dwelling than the Heart of a Man that is lodg'd in a Woman. There is no Point of the Compass to which they cannot turn, and by which they are not turn'd; and by one as well as another; for Motion not Method is their Occupation. To know this, and yet continue to be in Love, is to be made wise from the Dictates of Reason, and yet persevere to play the Fool by the force of Instinct.
(II. vi)[1]

This, for all its fancy, has a sobriety of tone, a faculty for sage yet witty generalisation, that marks out Mirabel as distinctive among lovers. And here is Mrs. Marwood speaking in soliloquy, when she has just learned that Mrs. Fainall has been Mirabel's mistress and that her own disappointed love for Mirabel is known to Foible the servant:

Indeed, Mrs. Engine, is it thus with you? Are you become a go-between of this Importance? Yes, I shall watch you. Why this Wench is the *Pass-par-toute*, a very Master-Key to every Body's strong Box. My Friend *Fainall*, have you carry'd it so swimmingly? I thought there was something in it; but it seems it's over with you. Your Loathing is not from a want of Appetite then, but from a Surfeit. Else you could never be so cool to fall from a Principal to be an Assistant; to procure for him! A Pattern of Generosity, that I confess. Well, Mr. *Fainall*, you have met with your Match.—O Man, Man! Woman, Woman! The Devil's an Ass: If I were a Painter, I would draw him like an Idiot, a Driveler with a Bib and Bells. Man shou'd have his Head and Horns, and Woman the rest of him. Poor simple Fiend! Madam *Marwood* has a Month's mind, but he can't abide her—'Twere better for him you had not been his Confessor in that Affair; without you could have kept his Counsel closer. I shall not prove another Pattern of Generosity—he has not oblig'd me to that with those Excesses of himself; and now I'll have none of him. Here comes the good Lady, panting ripe; with a Heart full of Hope, and a Head full of Care, like any Chymist upon the Day of Projection. (III. vii)

There is a strain of suppressed hysteria in this, with its ex-

---

[1] Quotations from Congreve are from *Comedies by William Congreve*, ed. Bonamy Dobrée (World's Classics, 1929).

clamations and its quick turns of thought, yet there is delicacy in the very railing and a kind of joy in the fanciful comparison of Lady Wishfort, as she thinks of an amorous encounter, to an alchemist on the brink of riches. Even in Congreve's first plays there is an evenness, a sureness, a delight in the flow of word and image, and at the same time a feeling of tension. Here is the middle-aged Heartwell in *The Old Bachelor*, as he finds himself against his will falling in love with Sylvia: he approaches her door and longs, and fears, to enter:

> Why whither in the Devil's Name am I a going now? Hum—let me think—Is not this *Sylvia's* House, the Cave of that Enchantress, and which consequently I ought to shun as I would Infection? To enter here, is to put on the envenom'd Shirt, to run into the Embraces of a Fever, and in some raving Fit, be led to plunge my self into that more consuming Fire, a Woman's Arms. Ha! well recollected, I will recover my Reason, and be gone. . . . Well, why do you not move? Feet do your Office—not one Inch; no, fore-gad I'm caught—There stands my North, and thither my Needle points—Now could I curse my self, yet cannot repent. O thou delicious, damn'd, dear, destructive Woman! S'death how the young Fellows will hoot me! I shall be the Jest of the Town: Nay in two Days, I expect to be Chronicled in Ditty, and sung in woeful Ballad, to the Tune of the superannuated Maidens Comfort, or the Batchelors Fall; and upon the third, I shall be hang'd in Effigie, pasted up for the exemplary Ornament of necessary Houses, and Coblers Stalls—Death, I can't think on't—I'll run into the Danger to lose the Apprehension.
>
> (III. ii)

Thus in the general fabric, the characterisation and the prose style of his plays, Congreve's fellow-poets could see an organised expression of the Restoration aim, a fusion of judgment and delight, with a greater range and a sharper sensitivity than English drama had known since the Civil War.

Yet the special quality of Congreve's work does not depend only on his bringing into a unity elements that can be found in drama of the early Restoration years. He was writing in the reign of William III, between the Revolution of 1688 and the accession of Anne. It was a time that saw the birth of new tendencies in the drama and in the society of London. In comedy, in particular, the last decade of the century was a time of important change. Colley Cibber presented his first play, *Love's Last Shift, or The Fool in Fashion*, in 1696, and the popularity of this led to the appearance a few months later of *Woman's Wit, or The Lady in Fashion*. Neither play can claim much of our respect, but there is no doubt that the earlier of the two ministered luckily to the taste of its time. The main action shows how Loveless, who deserted his wife Amanda after six months of marriage and fled abroad, returns to London penniless and is won back by Amanda through her device of pretending to be a stranger, conducting an intrigue with him and then appealing to his better nature. In his epilogue Cibber apologises for his hero's reformation, and says it has been contrived to please the ladies in the audience: the gentlemen should after all remember that 'He's lewd for above four acts.' But the popularity of the play showed the apology to be hardly needed. The people of the playhouse in the 1690s could still enjoy the bawdy phrase or situation, but they could evidently enjoy too the sententious utterance, the adaptation of the play's fable to the orthodox notion of how life should be lived. It is wrong to think of *Love's Last Shift* as merely consisting of four acts of commonplace Restoration intrigue followed by one act of sentiment: though Cibber enjoys the salacious aspect of Loveless's intrigue with his own wife, there is throughout the play a cultivation of the moral idea. Thus Young Worthy reproaches his brother's fiancée Hillaria for encouraging the attentions of the fop Sir Novelty Fashion:

> *Hil.*   The fool diverted me, and I gave him my hand, as I would lend my money, fan, or handkerchief, to a legerdemain, that I might see him play all his tricks over.
> *Y. Wor.*   O, madam, no jugler is so deceitful as a fop; for while you look his folly in the face, he steals away your reputation with more ease than the other picks your pocket.
> *Hil.*   Some fools indeed are dangerous.

177

Y. Wor. I grant you, your design is only to laugh at him; but that's more than he finds out: therefore you must expect he will tell the world another story; and 'tis ten to one but the consequence makes you repent your curiosity.

Hil. You speak like an oracle: I tremble at the thoughts on't.

Y. Wor. Here's one shall reconcile your fears—Brother, I have done your business: *Hillaria* is convinc'd of her indiscretion, and has a pardon ready, for your asking it.

El. Wor. She's the criminal; I have no occasion for it.

Y. Wor. See, she comes towards you; give her a civil word at least.

Hil. Mr. *Worthy*, I'll not be behind-hand in the acknowledgment I owe you: I freely confess my folly, and forgive your harsh construction of it: nay, I'll not condemn your want of good-nature, in not endeavouring (as your brother has done) by mild arguments to convince me of my error.

El. Wor. Now you vanquish me! I blush to be outdone in generous love! I am your slave, dispose of me as you please.

Hil. No more; from this hour be you the master of my actions and my heart.   (II. i)[1]

We should bear this passage in mind when we turn to Act II of *The Way of the World* and find Mirabel similarly reproaching Millamant for rejecting his company for that of Witwoud and Petulant:

Mira. . . . You had the leisure to entertain a Herd of Fools; Things who visit you from their excessive Idleness; bestowing on your Easiness that Time, which is the Incumbrance of their Lives. How can you find Delight in such Society? It is impossible they shou'd admire you, they are not capable: Or if they were, it shou'd be to you as a Mortification; for sure to please a Fool is some degree of Folly.

Milla. I please my self—Besides, sometimes to converse with Fools is for my Health.

Mira. Your Health! Is there a worse Disease than the Conversation of Fools?

Milla. Yes, the Vapours; Fools are Physick for it, next to *Assa-foetida.*

[1] Quotations from Cibber are from *The Dramatic Works of Colley Cibber, Esq.*, 5 vols. (London, 1777).

> *Mira.* You are not in a Course of Fools?
> *Milla.* *Mirabell*, if you persist in this offensive Freedom—
> you'll displease me—I think I must resolve after all, not to have
> you—We shan't agree. (II. v)

Of course, Congreve gives Millamant the victory of the last
word, but Mirabel's complaint is firmly made. Young
Worthy talked pompously of 'reputation'; Mirabel shows
impatience at a lack of aesthetic discrimination. Nevertheless, here *The Way of the World* includes within itself the
sententious strain of the newer comedy, yet with no breaking
of the play's fabric. Part of the secret is in Congreve's keying-
down of the sentiments and passions of his dramatis personae. Just as the thwarted love of Mrs. Marwood is never a
source of danger to the play's harmony, as that of Mrs.
Loveit is in Etherege's *The Man of Mode*, so here Mirabel's
seriousness makes its effect but Millamant's gaiety can
prevent his tone from becoming dominant. Mirabel indeed
must generalise at his peril when she is with him, as a
moment later in the same scene:

> *Mira.* I say that a Man may as soon make a Friend by his Wit,
> or a Fortune by his Honesty, as win a Woman with Plain-dealing
> and Sincerity.
> *Milla.* Sententious *Mirabell!* Prithee don't look with that
> violent and inflexible wise Face, like *Solomon* at the dividing of
> the Child in an old Tapestry Hanging. (II. v)

Paul and Miriam Mueschke have suggested that Cibber's
'vulgarity, casuistry, and irresponsible juggling with values'
may have strengthened Congreve's wish to write in a more
genuinely serious fashion.[1] That could well be true, even if
at the same time his comic range included something of the
newer mode. Cibber, moreover, in his flashy way could learn
something from Congreve. In his *The Careless Husband* (1704)
he produced one of the most popular plays in the sentimental
style, but he included in it the figure of Lady Betty Modish,

[1] *A New View of Congreve's Way of the World* (Ann Arbor, 1958),
p. 85.

a gay creature who only slowly and reluctantly yields to man and marriage. Like Hillaria in *Love's Last Shift*, she is charged with giving her time to fools rather than to men of sense, but Lady Betty is kept free from the solemnity with which the repentant Hillaria is exhibited. Sir Charles Easy, who reproaches her on behalf of his friend, presents a more vivacious picture than Young Worthy could manage:

> L. *Bet.*   . . . pray, sir, wherein can you charge me with breach of promise to my Lord?
>
> Sir *Char.*   Death, you won't deny it? How often, to piece up a quarrel, have you appointed him to visit you alone; and tho' you have promis'd to see no other company the whole day, when he was come, he has found you among the laugh of noisy fops, coquets, and coxcombs, dissolutely gay, while your full eyes ran o'er with transport of their flattery, and your own vain power of pleasing? How often, I say, have you been known to throw away, at least, four hours of your good-humour upon such wretches? and the minute they were gone, grew only dull to him, sunk into a distasteful spleen, complain'd you had talk'd yourself into the head-ach, and then indulg'd upon the dear delight of seeing him in pain: and by that time you had stretch'd, and gap'd him heartily out of patience, of a sudden most importantly remember you had outsat your appointment with my Lady *Fiddle-faddle*; and immediately order your coach to the Park?   (V. vii)

In his *Apology* Cibber says that he put the play aside after writing the first two acts, being 'in despair of having Justice done to the Character of Lady *Betty Modish*, by any one Woman, then among us', since Mrs. Verbruggen was in a declining state of health and Mrs. Bracegirdle was 'out of my Reach, and engag'd in another Company'.[1] He took it up again when he realised that Mrs. Oldfield had the skill and the temperament for the part. That he thought of Mrs. Bracegirdle in this matter suggests that Millamant was in his mind, and indeed the impress of that character is strongly on Lady Betty. The relation of Congreve to the emerging

---

[1] *An Apology for the Life of Mr. Colley Cibber*, 2nd ed. (London, 1740), p. 249.

new drama thus becomes more apparent in the linking of *Love's Last Shift*, *The Way of the World* and *The Careless Husband*. He viewed Cibber's success with disdain, describing *The Careless Husband* as a play 'which the ridiculous Town for the most part likes, but there are some that know better':[1] his opinion may have been sharpened by a measure of realisation that one element was common to Cibber's work and his own complex comedy.

Congreve's subdued seriousness of tone first becomes apparent in his second play, *The Double Dealer*. It is accompanied there by the appearance of two characters who scheme to overthrow the fortunes of the lovers Mellefont and Cynthia, as in *The Way of the World* Fainall and Mrs. Marwood scheme against Mirabel and Millamant. In the list of dramatis personae preceding *The Double Dealer* we find Maskwell labelled bluntly 'A Villain', and his actions and those of Lady Touchwood are presented, not comically, but as a manifestation of human evil. In this play Congreve's characteristic harmony is impaired: the villains and the lovers seem to belong to a different world from the Froths and the Plyants whose follies are displayed. It is noticeable too that Mellefont and Cynthia have not the wit and resiliency with which Congreve usually endows his lovers: he makes them more dependent on the gifts of Providence than on their own readiness to manipulate the skein of event. In *The Way of the World*, however, Fainall and Mrs. Marwood, at least in the first four acts, are admirably in tune: they are no mere villains, but people nourishing envy and a sense of wrong yet keeping it strictly under control. Here for example is Fainall speaking in prose with a strong nervous tension but with humour too, when he has learned that his wife has been Mirabel's mistress:

> Mrs. *Mar.* Well, how do you stand affected towards your Lady?
> *Fain.* Why faith I'm thinking of it.—Let me see—I am

---

[1] Quoted by F. Dorothy Senior, *The Life and Times of Colley Cibber* (New York, 1928), p. 58, n. 1.

Marry'd already; so that's over—My Wife has plaid the Jade
with me—Well, that's over too—I never lov'd her, or if I had,
why that wou'd have been over too by this time—Jealous of her
I cannot be, for I am certain; so there's an end of Jealousie.
Weary of her, I am and shall be—No, there's no end of that; No,
no, that were too much to hope. Thus far concerning my Repose.
Now for my Reputation,—As to my own, I Marry'd not for it;
so that's out of the Question.—And as to my Part in my Wife's—
Why she had parted with hers before; so bringing none to me,
she can take none from me; 'tis against all rule of Play, that I
should lose to one, who has not wherewithal to stake.

(III. xviii)

Only in the fifth act does *The Way of the World* fail to absorb
this villainous element into its harmony. There indeed Fainall
and Mrs. Marwood become melodramatic figures, and it
requires all Congreve's skill to make us accept the final stages
of the action when the thwarted villains have gone.

Now in Cibber's second play, *Woman's Wit, or The Lady in
Fashion*, we can see this element of villainy in the newer
comedy. There Leonora, a light-minded lady, is exposed in
her true colours by Longville, and in revenge she tries to ruin
his friendship with Lord Lovemore and his love for Olivia.
Her devices are many, and are successful till the very end of
Act V. When things go wrong for her, she cries 'Confusion!'
like any villain of melodrama, and indeed like Fainall in
the last act of *The Way of the World* when his wife's deed of
gift is produced. In the final speech of Cibber's play, the
virtuous Longville notes: 'we may observe that virtue ever
is the secret care of Providence'—just as *The Double Dealer*
ended with the assertion that mischief will always destroy
the villain that gives it birth. There is no question here of
one dramatist influencing another: *The Double Dealer* appeared three years before *Woman's Wit*, and neither play was
successful with the public: each appears to be making its
separate anticipation of the optimistic concern with hard-
pressed virtue that became prominent in eighteenth-century
comedy. But it was Congreve's achievement that his comedy
at its best—that is, in the first four acts of *The Way of the*

World—could include the villainy of a Fainall along with the wit of Millamant and the ludicrous exposure of Lady Wishfort, and through all these things combined could suggest an attitude of calm appraisal, made gentle through sympathy yet never irrational.

This attitude of Congreve is one fitting the end and summation of an age. He includes, we have seen, elements from the comic drama of forty years, deriving much from Etherege but something too from the newest comedy of his time. While not being old-fashioned, not a mere resurrectionist, he brings together all that is good, all that is truly exploratory, in Restoration comic writing. In the commendatory verses printed before *The Double Dealer*, as we have noted, Dryden praised him for having '*The Satire, Wit and Strength of Manly Wicherly*', and in his dedication of the play Congreve referred to his 'Satire' on Maskwell and Lady Touchwood. But most often he lacks the animus of the satirist. There is an affection in his presentation of Heartwell and Fondlewife in *The Old Bachelor*, of the Froths and the Plyants in *The Double Dealer*, of Tattle and Mrs. Frail in *Love for Love*, of Witwoud and Lady Wishfort in *The Way of the World*. The prologue to *Love for Love*, the most vigorous of his plays, comments on the decay of satire since Wycherley's time and adds that, though Congreve will here wield the lash, he '*hopes there's no Ill-manners in his Play.*' When he came to write *The Way of the World*, he could say in his prologue:

> Satire, he thinks, you ought not to expect;
> For so Reform'd a Town, who dares Correct?

Though this of course is a jibe at Collier, the lashing of abuses was, in fact, never much Congreve's concern: he recognises villainy and folly and will expose them, but apart from *The Double Dealer* the final impression each play gives is that the way of the Restoration world is amusing to watch and on the whole good enough. He holds up the mirror, composing his picture from what he sees reflected, and finds it pleasant. There is appraisal in his mind, but acceptance too. There is wit, and consideration. He had powerful links with the earliest years of Restoration drama—Dryden was his friend—but he came late enough to see the Restoration manners in perspective, to look at them with just a sufficient trace of the newer seriousness. Ten years later his comedy might have been sentimental and boneless. Coming when he does, he combines the wit and strength of earlier years with a stronger sympathy than Etherege and Wycherley knew. It was not merely that his satire was better-mannered than theirs: rather, he could mock at folly and pretension, yet realised the implications of human vulnerability.

In another respect, too, his position in time was fortunate. Earlier Restoration drama, though it had advanced a little towards the picture-frame stage, was still a frankly theatrical affair. The frequent use of rhyme in tragedy, surviving long after the short vogue of the wholly rhymed play, the elaborate conceits which Dryden and his contemporaries found natural to the dramatic style, the patterned speeches and situations in both tragedy and comedy, the love of fantastic incident and spectacular display, all marked off the traffic of the stage as belonging to a world apart. In the eighteenth century, however, the manner of drama became generally more homespun, less given to boldness of speech and spectacle in tragedy, less marked by the obvious contrivances of wit in comedy. Before Congreve's time the mode is high-pitched in tragedy, brittle in comedy; afterwards it generally became stolid and flat in both. Congreve has sufficient of the older manner to give his expression distinction and an air of authority, but he is close enough to the new manner for his scenes and characters to seem nearer to us, and therefore of more obvious concern, than those of Dryden, Etherege or Wycherley. This is not merely a matter of his prose style, though that too combines the virtues of fluency and study: it relates to the whole planning of the work. In the dedication of *The Double Dealer* Congreve was anxious to defend the play against the charge of implausibility: such characters as Maskwell and Lady Touchwood did exist, he averred, and it was common enough for a Mellefont to trust an untrustworthy friend. Moreover, apparently exception had been taken to Maskwell's revelation of his villainy in soliloquies. In view of the long-established use of the soliloquy in English drama, it is indeed significant that such a complaint was made. Congreve in his defence claims that, not the soliloquy, but the ill-advised use of it is to be blamed. It is better, he implies, that a dramatist should employ soliloquy than that he should implausibly introduce a confidant. A villain like Maskwell will not confide his villainy: if, therefore, we the audience are to know it, there must be soliloquy. Its use should, however, be hedged round with restrictions: the actor must be alone on the stage, or think himself so, and must on no account suggest an awareness of an audience:

> In such a Case therefore the Audience must observe, whether the Person upon the Stage takes any notice of them at all, or no. For if he supposes any one to be by, when he talks to himself, it is monstrous and ridiculous to the last degree. Nay, not only in this Case, but in any Part of a Play, if there is expressed any Knowledge of an Audience, it is insufferable.

We may question this as a theatrical principle, yet we can recognise that the element of naturalism in Congreve does contribute towards his special quality. He makes less use of the aside than we can find, for example, in Wycherley: after *The Old Bachelor*, in fact, he employs it rarely, and he has no scenes like that in *The Country Wife* where Alithea provides continual comments on Pinchwife's ill-advised warnings to Margery against the dangers of the town. The result in Congreve is that the scenes and characters, while exciting our interest and sympathy, seem yet to belong to a self-contained world, one that is sufficiently like our own to have plausibility but is at the same time disconnected, free. This sense of a combined plausibility and separateness gives to Congreve's dramatic scheme of things both fragility and wholeness. We are spectators of it only, but more fascinated spectators than if the plain manner of our world were mirrored there.

The very structure of the stage helped in this. There was a large front platform, with two permanent doors on either

G

side, and behind it there was painted scenery. Most of the action, most of the entries and exits, took place on the apron. Thus an antithesis was set up between a three-dimensional world in which the actors moved and a two-dimensional background which represented the world they were imagined as living in. Congreve's drama at its best depends on the tension between the individual figures and the society which imposes conventions, expectations, circumscription. Mirabel and Millamant, Mellefont and Cynthia, Valentine and Angelica, can assert themselves and win a measure of triumph; but that triumph must be fugitive, for the world remains fixed, unchangeable. The form of the Restoration stage, in its contrast of acting-area and pictorial scene, excellently objectified the basic element in Congreve's thought. It was not a theatre suitable for a great range of drama (Wycherley was perhaps never at home there), but Congreve's plays belong intimately to it. That its virtues were in part recognised can be seen from a passage in Cibber's *Apology*, where he compares it with the eighteenth-century stage:

It must be observ'd then, that the Area, or Platform of the old Stage, projected about four Foot forwarder, in a Semi-oval Figure, parallel to the Benches of the Pit; and that the former, lower Doors of Entrance for the Actors were brought down between the two foremost (and then only) Pilasters; in the Place of which Doors, now the two Stage-Boxes are fixt. That where the Doors of Entrance now are, there formerly stood two additional Side-Wings, in front to a full Set of Scenes, which had then almost a double Effect, in their Loftiness, and Magnificence.

By this Original Form, the usual Station of the Actors, in almost every Scene, was advanc'd at least ten Foot nearer to the Audience, than they now can be; because, not only from the Stage's being shorten'd, in front, but likewise from the additional Interposition of those Stage-Boxes, the Actors (in respect to the Spectators, that fill them) are kept so much more backward from the main Audience, than they us'd to be: But when the Actors were in Possession of that forwarder Space, to advance upon, the Voice was then more in the Centre of the House, so that the most distant Ear had scarce the least Doubt, or Difficulty in hearing what fell from the weakest Utterance: All Objects were thus drawn nearer to the Sense; every painted Scene was stronger; every grand Scene and Dance more extended; every rich, or fine-coloured Habit had a more lively lustre.[1]

Cibber praised the older stage for the greater ease of seeing and hearing that it offered, but he gives us a sense of what that stage must have meant for Congreve. Neither the homogeneous world of the early-seventeenth-century theatre nor the approach to the picture-frame that came in the eighteenth century could have constituted so remarkable an analogue to the tensions of his writing. In discussing the way in which Restoration comedy should be performed to-day, Norman N. Holland has urged a re-creation of the stage-dichotomy of its time:

> The set for a Restoration comedy ought to give a sense of the complex dialectic between the outer layer of sense-impressions and the 'solid' underlying core of personal and private life. The designer should give us a sense of the flatness of appearances and the roundness and depth of nature. One way of doing this— and I am sure there are many others—is to rely most heavily on flats and backdrops for the set. Only when absolutely necessary should a three-dimensional structure be permitted on the stage, for it will detract from the solidity that in Restoration comedy belongs only to the characters.[2]

It was supremely in Congreve that the characters achieved a recognisable solidity and, simultaneously, a full awareness of the thinness and rigidity of their environment.

Whatever the merits of his three earlier comedies—merits high above the general level of Restoration achievement— one is inevitably inclined to see them as preliminary sketches for *The Way of the World*. Yet, if they are preliminary sketches, each prepares the way in its own fashion. We have noted that *The Old Bachelor*, written when Congreve was twenty-three, has a harmony of atmosphere lacking in Etherege's

[1] *Apology*, pp. 339–40.

[2] *The First Modern Comedies: The Significance of Etherege, Wycherley and Congreve* (Cambridge, Mass., 1959), p. 237.

and Wycherley's first plays. It is remarkable, moreover, for the variety within its structure. There is the Etheregean quartet of lovers; there is the old bachelor of the title, with his love for Sylvia, the cast mistress; there is the cuckolding of the citizen Fondlewife; and there is the exposure of the blustering Sir Joseph Wittol. All is light and good-humoured, though of course not all is of an equal quality. Sir Joseph and his bully, Captain Bluffe, are crude, Shadwell-like figures, and the kicking of them by Sharper in Act III reminds us of the horse-play of the early Restoration years, seen for example in Dryden's *The Wild Gallant* (1663), where the rakish hero kicks the bawd and her companions because they have revealed their character to his admired Constance. The use of the forged letter, which makes Vainlove believe his Araminta more complying than he would have her, is a tiresome device, and indeed Congreve's plots are often marred by contrivance of this kind. On the other hand, the playwright shows an unusual perception of human frailty when he makes Vainlove unwilling to go through with either an intrigue or matrimony itself as soon as he finds the lady willing, and at the end of the play his future with Araminta is appropriately left unsettled. The scenes between Heartwell and Sylvia and between Fondlewife, his wife Laetitia and her quickly adroit lover Bellmore are in their way excellent: they go racily, with skill in the unexpected phrase and in the continuous action. In comparison with them, the scenes between the quartet of lovers are immature. The lovers can speak pleasantly and with a consistent grace, but they have not the precision of Etherege's lovers, and none of the four, even Vainlove, who is psychologically interesting, is sufficiently individualised: they stand mid-way between the Etheregean abstraction and the rounded character. The play's general atmosphere is suggested in Araminta's words:

> Nay come, I find we are growing serious, and then we are in great Danger of being dull. (II. vii)

Here seriousness and dullness are securely kept away.

That *The Double Dealer* is in a different strain is indicated by the quotation from Horace on its title-page: '*Interdum tamen, & vocem Comœdia tollit.*' Thus Congreve announces his more ambitious intention. The seriousness of purpose is seen not merely in the introduction of the villainous characters, Maskwell and Lady Touchwood, but in the manner of presentation of the two lovers Mellefont and Cynthia. As characters they lack individuality, but they stand for a man and a woman in love yet faced with the folly which seems inevitably to attend on their married acquaintances and faced too with the intrigues of the envious. In Act II, just after Lord and Lady Froth have given a fulsome exhibition of connubial affection, Cynthia doubts whether she and Mellefont are well-advised to proceed to marriage:

Mel.   You're thoughtful, *Cynthia?*
Cynt.   I'm thinking, tho' Marriage makes Man and Wife one Flesh, it leaves 'em still two Fools; and they become more conspicuous by setting off one another.
Mel.   That's only when two Fools meet, and their Follies are oppos'd.
Cynt.   Nay, I have known two Wits meet, and by the Opposition of their Wit, render themselves as ridiculous as Fools. 'Tis an odd Game we're going to Play at: What think you of drawing Stakes, and giving over in time?
Mel.   No, hang't, that's not endeavouring to win, because it's possible we may lose; since we have shuffled and cut, let's e'en turn up Trump now.
Cynt.   Then I find it's like Cards, if either of us have a good Hand it is an Accident of Fortune.
Mel.   No, Marriage is rather like a Game at Bowls, Fortune indeed makes the Match, and the two nearest, and sometimes the two farthest are together, but the Game depends intirely upon Judgement.
Cynt.   Still it is a Game, and consequently one of us must be a Loser.
Mel.   Not at all; only a friendly Trial of Skill, and the Winnings to be laid out in an Entertainment.   (II. iii)

As so often in Congreve, the force of the passage depends on its context: it has indeed a powerful ring when the Froths have just been exhibited to us. And the presentation of the lovers' dilemma goes further than this: Cynthia is aware that her powers of perception make life a more serious concern for her than it is for the Froths and the Plyants: she reaches a point in Act III where she can envy the fools, while knowing that her own wit cannot be laid aside:

'Tis not so hard to counterfeit Joy in the Depth of Affliction, as to dissemble Mirth in Company of Fools—Why should I call 'em Fools? The World thinks better of 'em; for these have Quality and Education, Wit and fine Conversation, are receiv'd and admir'd by the World—If not, they like and admire themselves—And why is not that true Wisdom, for 'tis Happiness: And for ought I know, we have misapply'd the Name all this while, and mistaken the Thing: Since

If Happiness in Self-content is plac'd,
The Wise are Wretched, and Fools only Bless'd.
(III. xii)

There is in this something of the elegiac note that can be heard in *The Way of the World*. If, however, Mellefont and Cynthia are better for the things they say than for the impression they make on us as imagined beings, Maskwell and Lady Touchwood are wholly mechanical pieces of villainy. Congreve was not capable of fully imagining the degree of malignity that he planned for them. He was to do better with Fainall and Mrs. Marwood, partly because their villainy is less and their excuse more powerful: envy and spite and the desire for revenge were of his world, but the fiercer, elemental passion was beyond its scope.

The play's title is an obvious echo of Wycherley, and Congreve certainly means here to expose villainy as well as folly. It has not the bite of Wycherley's best scenes, but Congreve, still aged only twenty-four, suggests that he has more to say than his predecessor. The play is at once an exposure and a programme. It puts in a plea for the rational virtue of Mellefont and Cynthia, it argues for good manners and judgment, it openly proclaims its earnestness without altogether losing a sense of its own small scope. *The Way of the World* was to combine much of this play's serious intent with the gaiety of the intervening *Love for Love*.

There is more vigour, heartier entertainment in *Love for Love* than in any other of Congreve's plays. There are no villains here, no serious threat—despite the many twists and turns of the plot—to the marriage-plans of Valentine and Angelica. The characters, for the first time in Congreve, are vivid: Foresight, the astrologer ready to believe in anything, even his wife's virtue; Sir Sampson Legend, never reluctant to consider trying for a new heir; the sailor Ben, with his remarkable quickness in the uptake; Tattle, the talker who always commits himself farther than he intends; Miss Prue, a juvenile cousin of Mrs. Pinchwife; the ladies Foresight and Frail, whose easy virtue is cheerfully worn—all are individually not outside Wycherley's scope. Wycherley, however, would not have been likely to put them all in a single play. For the secret of this play's popularity is largely in the diversity of its humour. Its contribution to Congreve's development consists in its use of strongly etched figures which are brought into harmony. In *The Way of the World* the characters have less visible outlines, but they retain something of the firmness that is apparent here. It is not only in the subordinate characters that this is found. As Tattle is superior to Brisk or Scandal to Careless, so Valentine and Angelica stand far more clearly before us than Mellefont and Cynthia. They are not so thoughtful as the earlier lovers were: there is almost nothing of the elegiac strain here, and Valentine's assumed madness leads, for the most part, only to grotesque humour. But they are self-assertive, they are not content to be imposed on as Mellefont and Cynthia were. It is Valentine who is responsible for much of the plot, till in the last scenes Angelica takes command and satisfies herself of Valentine's love.

Even in this play, however, we have a touch of the new sentiment. Valentine's gesture in abandoning his inheritance when he believes Angelica will marry his father, and Angelica's cry 'Generous *Valentine!*' when she discovers his intention, are equally unconvincing. We should like a hint that Valentine was fairly sure the sacrifice would not in the end be needed, and though we expect Angelica to like and indeed marry Valentine, we find her expression of admiration incompatible with her normal powers of perception. But that may be taking the end of the play too seriously. As in *The Double Dealer*, it is rounded off in a rather cavalier fashion, and certainly Congreve is too fond of elopements in disguise: the coming to terms of the lovers was in any event a settled thing, and the playwright did not concern himself unduly with its achievement. Nevertheless, the terms in which it is done remain appropriate to the comedy of sentiment. We are reminded a little of the noble gesturing at the end of *Love's Last Shift*, where the brothers Worthy and Sir William Wisewoud strive to outdo each other in generosity.

Five years separated the first productions of *Love for Love* and *The Way of the World*. In the interval Congreve had written a tragedy, *The Mourning Bride*, and clearly he came to his last comedy with deliberate steps. The prologue says:

He owns, with Toil, he wrought the following Scenes,

and the dedication describes the play as unsuited to the prevailing taste:

That it succeeded on the Stage, was almost beyond my Expectation; for but little of it was prepar'd for that general Taste which seems now to be predominant in the Pallats of our Audience.

That it failed to win much applause at its first performance is usually attributed to its defects in plotting, the thinness of the intrigue in the first four acts and the crowding of incident into the fifth. But not only is bustle wanting: the appreciation of *The Way of the World* needs a sensitivity to emotional overtone that is rare in an audience seeing a new play.[1] If for us this is the play that justifies all Restoration drama, for many of its contemporaries it was simply a less entertaining comedy than Vanbrugh's *The Relapse* (1696) or Farquhar's *The Constant Couple* (1699). Steele, in commendatory verses, echoes Congreve's motto for *The Double Dealer*:

> By your selected Scenes, and handsome Choice,
> Ennobled Comedy exalts her Voice;

but that was the judgment of a more than usually sympathetic spectator. To the discerning eye, indeed, there is more variety, more planning in *The Way of the World* than in *Love for Love*. The characters are less strongly etched, but their individualities are sure and the differences between them are brought out in the juxtaposition of scenes. It is no accident that the bargaining-scene between Mirabel and Millamant follows immediately after Sir Wilfull has approached Millamant with all the rustic embarrassment of the natural man, or that Mrs. Fainall, the cast mistress still mistress of herself, is placed side-by-side with Mrs. Marwood, the victim of thwarted inclinations. But the colours are subdued. Fainall does not gloat over his villainy as Maskwell did in *The Double Dealer*. Sir Wilfull brings the raw air of Shropshire into the drawing-room, but he does not nearly run away with the play as Ben did in *Love for Love*. Millamant is thoughtful like Cynthia, but not sententious, is in sure command of her speech but not given to pert repartee as on occasion Angelica is. Lady Wishfort is true to her name, but even her languishings and her vituperation have a smoothness and a comic elegance. Witwoud and Petulant are creatures of a summer's fancy, not earthbound in the way of Tattle. And Mirabel, for all his association with Mrs. Fainall, for all his pretended wooing of Lady Wishfort—which Congreve tactfully leaves outside the action of the play—appears not merely the wit but the

---

[1] The point was sharply made by Bonamy Dobrée in *Restoration Comedy* (Oxford, 1924), p. 140.

thoughtful man.[1] He is no victim of a sudden reformation, like Cibber's Loveless, but he has a plan for rational living, a plan in which Millamant constitutes a necessary part, a part to be kept in place. And for his own sake, he must see to it that she remains herself, no mere devoted complement to her husband. The bargaining-scenes that appear so frequently in seventeenth-century comedies, from the reign of Charles I right up to this year 1700, seem all anticipations of the meeting of Mirabel and Millamant in Act IV of this play: here Congreve developed the hesitations and the safeguards that he had sketched in the conversation of Mellefont and Cynthia: Restoration comedy had been principally concerned with sex-relations, and here Congreve gave to the dying age a satisfying expression of the attitude it had been fumbling for.

But the final impression the play leaves is perhaps one of suppressed melancholy. At the end of Etherege's *The Man of Mode*, the fortunate Harriet, secure of Dorimant's love, joins a little grossly in the jeering at the cast-off Mrs. Loveit. When Millamant speaks to Mrs. Marwood, there is an echo of this, but Congreve's raillery is altogether less pert: indeed, at the end of this encounter the mocking tone becomes subdued. To Mrs. Marwood's protest that she hates Mirabel, Millamant replies:

> O Madam, why so do I—And yet the Creature loves me, ha, ha, ha. How can one forbear laughing to think of it—I am a Sybil if I am not amaz'd to think what he can see in me. I'll take my Death, I think you are handsomer—and within a Year or two as young.—If you cou'd but stay for me, I shou'd overtake you—But that cannot be—Well, that Thought makes me melancholick—Now I'll be sad. (III. xi)

With Congreve we do not forget the passage of time: its ruins may amuse us, as in Lady Wishfort, but even in her

---

[1] Congreve, while keeping our sympathy with Mirabel, recognises a considerable range of conduct in him. Norman N. Holland, in an otherwise most perceptive study of the play, simplifies notably in labelling Mirabel 'a good clever man' (*op. cit.*, p. 160).

presentation there is a reminder that she is not alone. Sometimes the elegiac note seems to be communicated through the cadence, as when Mirabel speaks in Act II:

> An old Woman's Appetite is deprav'd like that of a Girl—
> 'Tis the Green-Sickness of a second Childhood; and like the
> faint Offer of a latter Spring, serves but to usher in the Fall;
> and withers in an affected Bloom. (II. iii)

And in Act I, before we have met Millamant, Fainall and Mirabel are discussing her and this exchange ensues:

> *Fain.* For a passionate Lover, methinks you are a Man somewhat too discerning in the Failings of your Mistress.
>
> *Mira.* And for a discerning Man, somewhat too passionate a Lover; for I like her with all her Faults; nay, like her for her Faults. Her Follies are so natural, or so artful, that they become her; and those Affectations which in another Woman wou'd be odious, serve but to make her more agreeable. I'll tell thee, *Fainall*, she once us'd me with that Insolence, that in Revenge I took her to pieces; sifted her, and separated her Failings; I study'd 'em, and got 'em by Rote. The Catalogue was so large, that I was not without Hopes, one Day or other to hate her heartily: To which end I so us'd my self to think of 'em, that at length, contrary to my Design and Expectation, they gave me ev'ry Hour less and less Disturbance; 'till in a few Days it became habitual to me, to remember 'em without being displeas'd. They are now grown as familiar to me as my own Frailties; and in all probability in a little time longer I shall like 'em as well.
>
> *Fain.* Marry her, marry her; be half as well acquainted with her Charms, as you are with her Defects, and my Life on't, you are your own Man again.
>
> *Mira.* Say you so?
>
> *Fain.* Ay, ay, I have Experience: I have a Wife, and so forth.
> (I. iii.)

Mirabel lovingly considers the twists and turns of his mistress's mind and delights in his own subjection: there is wisdom in the comic presentation as well as shrewdness and good phrasing. And the last words in the scene are given to the man of experience, who has tried marriage for himself. The acid is there, the shrug that goes with Congreve's wit. Bonamy Dobrée has noted that many of Congreve's set speeches are just not metrical, that a regular iambic beat is never quite established.[1] This may be the formal analogue to that undertone of melancholy which runs through the same speeches, giving to *The Way of the World* in particular an added dimension within which the characters move as symbols: there their separateness vanishes and they become aspects of a single humanity. Congreve, bringing Restoration comedy into clear focus, brought it also, for a brief moment, into a larger world.

---

[1] *Comedies by William Congreve*, ed. Bonamy Dobrée (World's Classics, 1929), p. xxiv.

Chapter 10

ON SEEING A PLAY

OUR concern here is with the experience that we have when we form part of a theatre-audience and are watching a play being acted in our presence, with the inevitable corollary of critical judgment. However relaxed, however indifferent we may be, our minds will nevertheless be forming an opinion of what we are offered. Through our response in laughter or in the manifest degree of attention that we give, we shall in a measure be making public the results of that critical process even while we do not know as a whole the play we are responding to. Indeed, it is more complicated than this. Every actor knows both that no two performances of a play are identical and that the difference between performances is in part caused by the differing responses from the audience. So our process of critical judgment in the theatre has, in a unique way, the power to modify the work that is being criticised. When we read a novel or a poem, a single public thing exists for us to come into accord with, to make up our minds about. It is true that, in the case of a long-established piece of writing, we read in the light of an existing and growing tradition of critical response. Somerset Maugham has remarked: 'I have a notion that the odes of Keats are more beautiful than when he wrote them.'[1] But when we go to see a new play, the performance as we shall see it has not yet achieved any full condition of existence, and the details of its character will be in part determined by our own behaviour while it is being acted. We, and not merely our predecessors as readers, exert an effect on the nature of the work. Certainly

[1] *The Summing Up*, London, reprinted 1940, p. 307.

in the subsequent exchange of ideas with one another we shall refer to a play as an entity without regard to its varying nature from night to night, without regard to our own part in shaping it. This perpetually shifting character of dramatic work must, however, be returned to later. What I am primarily concerned with for the moment is that we are impelled and invited to display a view of the work during the very time of impact. Even if we keep as neutral an appearance as possible, trying to avoid laughter or any sign of emotion or of profound engagement, we may well be asked to comment during the intervals in the performance.

Now this is the very opposite of what is recognised as good critical practice. We are for ever urging upon ourselves the need to submit to a piece of writing under consideration before committing ourselves to an opinion. Although we recognise that in fact we shall be always aware of a growing critical attitude during our reading, we try to prevent this from taking too hard an outline—not only until our reading is completed but until we have had time for second thoughts, time to consider the writing coldly (for coldness and warmth are equally necessary in a total judgment), to relate it to its aims and to writings of a similar Kind, and to reach some conclusions concerning the status of the Kind itself. We should, moreover, be ready to contemplate the possibility that our view will be substantially changed with the mere passing of time: it would be a strange reader who had not experienced this happening, though we have no guarantee that the later judgment is the superior one. In his 'Digression concerning Criticks' in *A Tale of a Tub*, Swift draws an ironic picture of a '*True Modern Critick*', bestowing on the creature exactly those features that Swift considers inimical to true criticism. So he urges his readers to observe, as a first principle,

> That *Criticism*, contrary to all other Faculties of the Intellect, is ever held the truest and best, when it is the very *first* Result of the *Critick's* Mind: As Fowlers reckon the first aim for the surest,

and seldom fail of missing the Mark, if they stay not for a Second.[1]

Premature judgment, in fact, is on a level with duck-shooting, demanding some dexterity with the voice or pen but showing no more regard for the writing criticised, attaching no more importance to the kind of thing it is, than will operate for Hemingway's Colonel Cantwell, crouching stiffly among the reeds before dawn. A hundred years later than Swift we find Wordsworth, in his different manner, expressing a similar view. In the Preface to the *Lyrical Ballads* of 1800 he was aware that the poems he was offering the reader ran counter to common expectation, that the immediate response would be for that reason almost certainly antagonistic, and he was doubtless aware too of the gulf between his aim and his achievement in some of the poems. So he did not want the poems condemned because the principles were new, he did not want the principles condemned because in some instances they were not successfully embodied in the poetry. More than most writers, therefore, he was anxious for his readers not to make a snap judgment. So, a little portentously, he reminds them of the difficulty of the critic's task, the need for comparative judgment, the need for patience before a critical view is allowed to assume finality:

> an *accurate* taste in poetry, and in all other arts, as Sir Joshua Reynolds has observed, is an *acquired* talent, which can only be produced by thought and a long-continued intercourse with the best models of composition. This is mentioned, not with so ridiculous a purpose as to prevent the most inexperienced reader from judging for himself (I have already said that I wish him to judge for himself); but merely to temper the rashness of decision, and to suggest that, if poetry be a subject on which much time has not been bestowed, the judgment may be erroneous; and that, in many cases, it necessarily will be so.[2]

[1] *Swift: Gulliver's Travels and Selected Writings in Prose and Verse*, ed. John Hayward, London, 1934, p. 316.

[2] *Wordsworth's Literary Criticism*, ed. Nowell C. Smith, London, 1905, p. 39.

But are we likely to 'temper the rashness of decision' when during the first interval in a dramatic performance our companion turns to us and says: 'He just doesn't know how people live', or 'A bit pretentious, don't you think?', or simply 'For God's sake, let's go to the bar'? Strindberg in his *Open Letters to the Intimate Theater* commented impatiently on the difficulty of maintaining a proper response to a performance when the theatre allows us to escape during an interval:

> The drawbacks of letting the audience out to imbibe strong drinks in the middle of the drama are well known. The mood is destroyed by talk, the transported spirit loses its flexibility and becomes conscious of what should remain unconscious; the illusion the drama wanted to give cannot be sustained, but the theatergoer who has been half carried away is awakened to utterly banal reflections, or he reads the evening paper, talks about other things with acquaintances he meets at the bar; he is distracted, the threads in the play are cut, the development of the action is forgotten, and in a completely different mood he returns to his seat to try in vain to pick up what he had left.[1]

Yet perhaps Strindberg has missed the main point here. There is a case for the interval, the pause in which we assimilate while exchanging greetings with those we meet or attending to our companion for the evening: we may be readier afterwards to encounter the dramatist's further exploration. The real danger of the interval is that we may commit ourselves to a particular judgment which we ought to discard when the play is over. And during the performance itself there will often be a strong inclination to prepare a critical comment suitable for utterance as soon as the final curtain has fallen: Tom Stoppard in *The Real Inspector Hound*, acted in London in 1968, made broad fun of this in relation to two professional critics preparing their notices as they watched. Even if we are alone and pay no attention to the comments of our neighbours, even if we try to make

---

[1] *Open Letters to the Intimate Theater*, tr. and ed. Walter Johnson, Seattle and London, n.d., p. 20.

our responses to the first act as provisional as possible, an anticipatory view of the total performance will be forming itself in our mind. And, though we may not listen to a word anyone else in the auditorium says, we shall certainly be aware that other anticipatory judgments are being formed. We shall probably be conscious either of going along with, or going against, the general tide of response: in either instance our purely anticipatory judgment will be given a harder outline than is the case when we are reading a book and we pause between chapters.

A further problem arises in the criticism of a play, and that is to determine what we are criticising. There are the performances of the actors, the contributions of the scenery and costume designers and the lighting-technicians, and the planning and control for which the director is responsible; there is, moreover, the work of the author, who has imagined the action and the characters and the words. At an ideally satisfying performance, of course, we should on consideration give our approval to each of these in turn. In practice we recognise a higher degree of merit in some features than in others. Yet that should be only after analysis, when we have been able to decide whether, for example, the playing of a particular part in a certain way was due to a failure in the performer's skill or to a deliberate plan to relate that part to the total effect. We may then decide that what appeared at first to be a poor performance was actually a contribution to the play's success, or that the failure was a real one but was due, not to a lack of skill in the actor, but to a mistake in interpretation or execution for which the director was responsible. But in the theatre we are invited to apportion praise or blame at once. The actors appear singly or in small groups to take their curtain-calls, and thus to learn approximately the degree to which they have pleased us. An audience inevitably makes a snap distinction, as rapid as any made by one of Swift's duck-shooters, between the varying merits of different aspects of a performance they have just seen. Some years ago *The Times* gave a report on the West Berlin production of John Whiting's *The Devils*. It recorded that, while the production as a production was applauded, the author was booed when he appeared on the stage, at least by a large section of the audience, who thus wished to show that, while paying tribute to the director, they had no regard for the writing he had given his time to. This was an unexpected incident, for Whiting's work, as it was played in London, seemed massive and eloquent though over-assertive: I am referring to it because it is typical of a kind of audience-response that needs attention. Here was an audience deciding in a moment that the director was blameless and indeed worthy of applause, while the author deserved humiliation. Yet a play as presented may take on certain properties, or even a total implication, that come from the director's rather than the author's hand. This, of course, is particularly the case when a play is being performed in translation, and is thus in any event somewhat removed from the author's imagining. I did not see the production in Berlin. For all I know, the director was wholly faithful to Whiting's intention. The point is, however, that the audience could not know whether he was. Moreover, a director who is faithful to his author (as of course he should be, though we have already noted examples of deliberate distortion[1]) has for the occasion of the performance entered into a condition of personal involvement with the text the author has given him. There should be no question of his working independently of the text, though he may of course illuminate that text by the employment of a theatrical means that the author had not thought of. If the text is bad, the director may show ingenuity, but he should not be praised for ingenuity in a bad cause. If the author is booed, the director should be booed too: it is he who has made the play available for inspection, given it the life of performance, brought out the lines of its composition and the texture of its material. When a performance has gone wrong, a director should not be told by his friends 'You were wasted on that play', but rather 'You wasted yourself on it'.

[1] See above, '*Catharsis* in English Renaissance Drama', p. 125.

In the criticism of drama we are both immediately and ultimately concerned with the judgment of a total performance. In the interim we may select this or that feature for a different degree of praise or blame, but it is the total judgment that matters, the decision between acceptance and rejection. Of course, there are many degrees of acceptance, and in deciding between these degrees—as we should —we shall take into account the measure of success won by the various contributors to the performance. But we should try to prevent our judgment, particularly a premature judgment, of a part from getting in the way of our judgment of the whole: the practice of the theatre encourages us to let that happen.

The situation is more complicated when we are watching a play whose composition is separated from us by a considerable stretch of time. In English-speaking countries in the case of the better-known plays of Shakespeare the problem is not so acute, for these plays have a fairly long acting-tradition and they are widely known through the reading of the text. There is therefore a ready-made expectation in the audience, many of whom will be seeing familiar figures and a familiar action brought once again to life. This has its own dangers, for on the one hand the audience may have a conservative attachment to the play as they have long known and perhaps mistakenly seen it, and on the other hand the director, wishing to break through the lethargy of custom, will be inclined to labour after a new mode of presentation, a mode which may be imposed quite arbitrarily on the play. Nevertheless, the special problems hitherto under discussion will apply here only in a slighter degree. We already have a total view of the play, which may be modified, may be enriched, by the new performance; but we are not finding our way in unknown country. And we shall be in a much more advantageous position to distinguish between the contribution of the dramatist on the one side, a contribution already known to us, and the contributions of director and actors and the rest. But the problem of judgment is at once more considerable when we turn to the lesser-known plays of Shakespeare, those which do not get often to the stage or are even much studied. Here for the majority of the audience the play will be new, with all the problems associated with newness, and at the same time it will be old, it will be a play which the author has manifestly addressed to a theatre-public remote from us. We shall have the sensation of eavesdropping on a theatre-experience of the past, with whose conventions, with whose peculiarities of prejudice and concernment, we must try to get in tune.

Even so, such a play has a certain immediate currency because it is Shakespeare's: a measure of expectation is with us, a sense of what the whole will probably be like. It is the non-Shakespearian drama of the past that an English-speaking audience finds special difficulty with. It is only recently, after a long interval, that major plays by Shakespeare's contemporaries and successors have begun to come with some regularity to the stage: in this respect the countries of continental Europe, in their treatment of a national repertory, have a marked advantage over the United Kingdom and North America. Few of us can expect to see Dryden's *Aureng-Zebe* or, unless we are especially fortunate, even Marlowe's *Tamburlaine* on the stage. It is noticeable that newspaper critics in London commonly write with a resentful embarrassment when a play by one of Shakespeare's contemporaries is produced. They do not know the play in advance, though texts for reading can be had; they are in the dark about current scholarly views on the worth of the play, though scholars may more than frequently write on the subject; they have a vague expectation that the play will be, or should be, like Shakespeare's though not so good. In such a frame of mind they discuss Middleton and Rowley's *The Changeling* or Ford and Dekker's *The Witch of Edmonton* or even Webster's *The Duchess of Malfi*, and have some difficulty in confronting the expectation with the fact. If indeed Webster is in question, they may remember a remark of Shaw's that associated him with the Chamber of Horrors in a waxwork museum and utilise it for one sentence of the review that has to be written. More damaging is it that the Jacobeans are treated as if all were alike in method and outlook: *The Changeling* has been criticised, when produced in London, as being inferior Webster. But, though such halting criticism may make us impatient, we should remember that a performance of a little known play of the fairly remote past presents a formidable critical problem. If it is a play of a high order, it will of course have something substantial to say to us now, but the manner of the saying, the details of the statement, will be conditioned by the circumstances of its time. In order to respond in a satis-factory fashion, we must make use of our historical imagina-tion, must try to see the play, in our mind's eye, as it appear-ed to its first audience, while at the same time with our physical eye we are seeing it acted on a modern stage by actors and actresses who live along with the audience in the twentieth century. It is also fair to remark that within the last decade there has been a noticeable improvement both in the knowledge brought to the theatre and in the readiness to see differences between one writer and another. This has been owed in substantial measure to the greater availability of seventeenth-century plays on the stage.

The theatre is always a place that demands a doubleness of response. We see a play and know it is a play, put on the stage after the author has written it, after the actors have rehearsed it, after the designers and technicians have done their work. At the same time we have a sense of immediacy, of being in the presence of people actually going through the experiences presented. Again, we know the ending is planned, inevitable; simultaneously we feel hope or fear for the characters. That remains true, however strenuously the dramatist and his director work against the idea of an easy 'illusion', a ready identification with the character's plight. We see the exaggerations of comic behaviour, the fortunate coincidences of romance, the unblurred pattern of the tragic procession; yet for a moment we feel that these things belong in our world, when of course our world lacks those simplifications, that degree of abstraction. In watching a play of the past we have to accept a further doubleness: we must have both a sense that Middleton or Webster or Ford, for example, is speaking directly to us, and a sense that we are finding our way back to the early seventeenth century and forming an understanding of what their utterances meant to their contemporaries.

The most complicated case I remember of the interrelation of past and present was one deliberately contrived. The occasion was the London performance of Orson Welles's dramatisation of *Moby Dick* in 1955. The play was acted as if it were a rehearsal in a provincial American theatre in the late nineteenth century. Simultaneously we were in the Duke of York's Theatre, London, in the imaginary American theatre where the rehearsal was taking place, in the same American theatre during an actual performance for which we were seeing the preparation (though at the end it was hinted that the performance would never take place), and on the deck of the stage-version of Melville's Pequod, which we remembered from the novel. Thus we were at four removes from the world of the novel, and the distancing made it all the easier to accept the necessarily stylised presentation of the action along with the translation to the stage of Melville's elemental figures. This may give a hint as to the best way of staging plays of the past, though I am far from suggesting that the production of an old play should take the same degree of freedom as is fully permissible in the adaptation of a novel. The director who brings a play of the past before his audience should not be afraid of being markedly different in manner from the contemporary theatre. The more obviously he transports us into a theatre of a different sort, the more readily we shall accept the terms in which the dramatic statement is made. It is a mistake to try to make an old dramatist look like a new one, just as it is a mistake to try to present a verse-drama in a prose disguise. On that condition of things rests part of the case for the use of an open-stage theatre for the performance of Elizabethan and Jacobean plays. An old dramatist's words and imaginings are more likely to seem 'new' if we do not try to make his work look 'contemporary'.

From this we may approach a problem of a different sort, a problem that lies near the heart of the dramatic experience. Drama is a communal rite. In its early history this was obviously so. The Athenians went to the theatre of Dionysus three times a year; the whole town, almost, was there, and it was watching not merely a poet's presentation of a usually traditional story but a ceremony in honour of the god; the priests had reserved seats (like the canons in a cathedral), and the god himself was present in effigy. The people of the Middle Ages went to see performances that showed them the history of the world as the Bible had told it, with the frequent addition of a glimpsed Last Judgment: the lesson was that of the church, made more vivid through impersonation and action, and the people at large were to respond to a shared doctrine. In both these instances attendance was itself an act of worship, although in later days in Athens some of the worshippers doubtless had personal reservations, personal ways of interpreting the myths, and the dramatists, too, might give to the gods only an oblique recognition. In a modern theatre, however,—that is, from the late sixteenth century onwards—the secularised drama has not been committed to the idea of proclaiming religious doctrine. Nevertheless, by joining a theatre-audience one is joining a group, coming into relationship with the dominating notions of that group. The audience, as we have seen, will itself affect the performance: their presence will be directly felt by the players, they are not outside the play. The theatre thus remains a place where group-consciousness, group-ideas, persist, where an audience will go with the expectation that their own established way of seeing things will be given authoritative expression on the stage. Fortunately, they do not regularly receive what they anticipate.

Yet the idea is implicit in Aristotle and explicit in Horace. In *The Poetics* Aristotle assumes that there are certain fundamental ideas concerning human conduct which the audience as a whole will inevitably share. On considering the possibility of a tragedy which showed an extremely bad man passing from happiness to misery, he comments:

> Such a story . . . will not move us to either pity or fear; pity is occasioned by undeserved misfortune, and fear by that of one like ourselves; so that there will be nothing either piteous or fear-inspiring in the situation.[1]

That is, he assumes a steadfast line of distinction between the envisaged spectators or readers and the 'extremely bad man', and he assumes that we too can make an easy distinction between deserved and undeserved misfortune. Now these distinctions are commonly made, in idle gossip and in newspaper articles and in the speeches of politicians: they constitute part of the automatically accepted, unscrutinised body of doctrine which most men assume to be true. And this implication of a minimal common attitude which can be expected in a theatre-audience, and which can indeed be discerned as linking the present with the past, is given overt expression in the *Ars Poetica* of Horace, when he speaks of the tragic chorus as voicing traditional wisdom:

> The Chorus must back the good and give sage counsel; must control the passionate and cherish those that fear to do evil; it must praise the thrifty meal, the blessings of justice, the laws, and Peace with her unbarred gates. It will respect confidences and implore heaven that prosperity may revisit the miserable and quit the proud.[2]

That is, the chorus will give utterance to the generally held aspirations of men, begging the gods that those who adhere to traditional virtue shall be rewarded. If we completely accepted this, drama would always be on the side of tradition, of convention, always ready to offer tribute to values that were becoming outworn. And, of course, in a measure this is true. Shaw in Act II of *You Never Can Tell* made one of his characters assure us that old-fashioned ideas always

---

[1] Chapter XIII. The translation is Bywater's.

[2] *Horace on the Art of Poetry*, ed. E. H. Blakeney, London, 1928, p. 49.

seemed up-to-date in the theatre. Yet, if an audience commonly expects to be told either the things it already believes or the things it is beginning to entertain the possibility of believing, it is the habit of the major dramatist to run counter to this expectation. We know well enough how irreverent was Euripides' attitude to the gods to whom he could give no easy belief, and even Aeschylus, commonly thought of as a man of correct principles, did not hesitate to show the petty malice of Apollo's vengeance in the *Agamemnon* and his outdated notion of justice in the *Oresteia* as a whole. Indeed we have seen how H. D. F. Kitto has argued that the *Oresteia* dramatises a turning-point in the conception of justice, moving away from a personally con-ducted blood-feud to the civic dispensation of law and equity. This trilogy, in celebrating the recent establishment of the court of the Areopagus, was paying tribute to a social revolution. Its doctrine was not new, but it was still challeng-ing, still at odds with deep-seated prejudice; and it put, we may say, the gods in their place: they, or what they ultimately stood for, were still revered, but man's responsi-bility to man was the matter underlined.[1]

In recent times we have seen obvious examples of a dramatist administering shocks to his audience's prejudices. When Shaw's first play, *Widowers' Houses*, had been acted, he wrote in the Preface to *Plays Unpleasant*: 'I had not achieved a success; but I had provoked an uproar; and the sensation was so agreeable that I resolved to try again.'[2] Indeed he had provoked an uproar: he had shown nice people living on the rents from slum-property, sometimes without knowing it, and he had shown a personable and expensively-brought-up young woman striking a servant and erotically pursuing a young man. His audience knew well enough that such things happened, but they did not expect to be reminded of them in the theatre. Moreover, in his sub-

---

[1] See above, '*Catharsis* in English Renaissance Drama', p. 127.

[2] *Plays: Pleasant and Unpleasant. The First Volume, containing the three Unpleasant Plays*, London, reprinted 1929, p. xii.

sequent plays Shaw preached socialism to solid citizens who had been induced to pay for admission. He delighted in a battle with his audience, at first confounding their expectation of what the play's pattern would be, and then administering doctrine which he knew they did not adhere to. Yet, and here is a paradox, they had the sense of being for a moment in tune with the current movement of ideas. They did not approve; they shrank within themselves to keep out the chill; yet they had at the same time a curious sense of solidarity with the growing number of people outside the theatre whose subversive notions were being mirrored by Shaw. Moreover, in a sense Shaw was with his audience. He gave them witty dialogue, traditional images of pretence and hypocrisy, and dashing young men and women who found their way to the traditional bed even if their path to it was an unusual one. A dramatist commonly links in this way a satisfying of the audience's expectations with a defiance of them, as is quite variously apparent in some plays of the last few years—for example, in Beckett's *Waiting for Godot*, Shelagh Delaney's *A Taste of Honey*, Arnold Wesker's *Trilogy*. There need be nothing 'insincere' in this linkage. The dramatists genuinely share in certain ways of thought and feeling that their audiences have, as Beckett values companionship and imagines a revelation that does not, cannot come, Shelagh Delaney gets a personal excitement from contemplating the unbreakable solitude of the girl who is her play's centre, Wesker makes every effort to preserve a belief in the holiness of the heart's affections: such ideas and feelings are common to playwright and audience. Indeed, if there were a total or near-total difference between the writers' outlooks and predispositions and those of the audiences that the plays have attracted, these writers would probably not write for the theatre or would not achieve stage-presentation. But, along with these things thus shared with the audience, Beckett shows his sometimes good men in a universe purely terrifying, Shelagh Delaney draws an often brutal cartoon of normality in the lower reaches of an affluent society, Wesker pleads for social revolution.

Many members of their audiences have the uneasy but interesting sensation of swinging between the acceptable and the unacceptable.

If this is true of Euripides and even Aeschylus on the one hand and of twentieth-century drama on the other, we should not be surprised if it is also true of the drama in the stretch of time that links the two periods. Or, rather, we should not be surprised if it happens whenever the drama occupies a prominent place in a nation's cultural life, as it did in seventeenth-century England. We know that certain Restoration plays provided a shock to their society, plays that included Wycherley's *The Country Wife* and Dryden's *Mr. Limberham, or The Kind Keeper.* Yet we are often ready to credit the notion that Elizabethan and Jacobean plays faithfully and invariably echoed the commonplaces of their time, as if not only Shakespeare (of whom it seems highly doubtful) but also Marlowe and Webster and the rest were content to enunciate the official doctrine of Canterbury and Whitehall. Here, however, we may look at a particular example of the way in which audience-expectation could be deliberately played upon—half-satisfied, half-frustrated—in the drama of that time. It may suggest that in the total statements of the plays we should look for a corresponding complexity.

Both Aristotle and Horace had insisted that the characters of a play should be true to the type they belonged to: a woman, for example, should not, according to Aristotle, be manly or clever;[1] an old man, according to Horace, must speak in the accents expected of age.[2] This doctrine was carried to an extreme by Thomas Rymer at the end of the seventeenth century. In *The Tragedies of the Last Age* of 1677 he criticised Beaumont and Fletcher for their presentation of Evadne in *The Maid's Tragedy* because in her, he declared, they contradicted the general notion of Woman:

[1] *The Poetics*, Chapter XV.
[2] *Ars Poetica*, ll. 114–16.

Now Nature knows nothing in the *manners* which so properly and particularly distinguishes woman as doth her modesty. Consonant therefore to our principles and Poetical, is what some writers of Natural History have reported; that women when drowned swim with their faces downwards, though men on the contrary.

Tragedy cannot represent a woman without modesty as natural and essential to her.[1]

And Evadne had indeed very little modesty in her conduct and speech. Similarly in *A Short View of Tragedy* of 1693 he attacked Shakespeare for presenting Iago in a way that was out of accord with the general notion of a Soldier:

to entertain the Audience with something new and surprising, against common sense, and Nature, he would pass upon us a close, dissembling, false, insinuating rascal, instead of an open-hearted, frank, plain-dealing Souldier, a character constantly worn by them for some thousands of years in the World.[2]

Now Rymer knew well enough that there were immodest women and dissembling soldiers in the world, but he believed that an audience expected that in a tragedy a woman should show modesty and a soldier plain-dealing: there was a common belief that such qualities belonged to the types as types, whatever eccentricity might be found in an individual and mildly exposed for correction's sake in a comedy. And in that surely he was right. We have noticed that one of the ways in which Shaw shocked his audience in *Widowers' Houses* was by showing his attractive young lady forgetting her upbringing (with the nature generally assumed to go with it) and striking her servant: though Shaw was writing comedy, he went beyond its acceptable limits here. Similarly Beaumont and Fletcher administered a shock in the second act of *The Maid's Tragedy* when the bride Evadne

---

[1] *The Critical Works of Thomas Rymer*, ed. C. A. Zimansky, New Haven, 1956, p. 64.

[2] *Ibid.*, p. 135.

suddenly reveals to her husband Amintor that she is the
King's mistress and that her marriage is to be nothing but
a cloak. The audience is obviously intended to be disturbed
by the violence of her mocking exclamation: 'A maidenhead,
Amintor, At my years?' So, too, in *Othello* Shakespeare
plays on the expectations of the audience in relation to
Iago. Not only is he a soldier, but he is a man whom everyone but Emilia sees as bearing the traditional character
of the Soldier: he is always 'Honest Iago', he is the man
to be trusted as Desdemona's escort to Cyprus, as the
man who will give a true report of the night's brawl in which
his superior officer Cassio was involved, even as the man
who will tell the truth in accusing another man's wife. As
C. A. Zimansky has put it in his edition of Rymer's critical
writings,

> Rymer, as so often, has pointed out the problem without giving
> the right answer. One can point out that it is improbable for a
> soldier to be so coldly calculating a villain, that the very
> improbability enables Iago to impose on Othello, and that his
> success depends partly on the idea of the typical soldier
> that Rymer holds—everyone believes that he has the soldier's
> qualities of simplicity and forthright honesty. So the very idea
> of decorum that Rymer upholds is actually in the play, and its
> violation allows the tragic action.[1]

But in this tragedy we have a second example of the manipulation of the audience's expectations. Othello is a black
Moor, and there was an Elizabethan dramatic tradition
which associated villainy with such a character-type: Aaron
in *Titus Andronicus* (though with a defiantly human touch
characteristic of Shakespeare) is only one example from a
whole series. And there is in *Othello* an element which
associates the hero with the earlier dramatic portraits of
black Moors. Although Brabantio has been on good terms
with Othello, he is ready to impute witchcraft to him as soon
as he hears of the elopement. Moreover, Othello does on
occasion act and speak with the violence that belongs to the

---

[1] *Ibid.*, p. xxviii.

type which he would in the theatre be recognised as belonging to.[1] Nevertheless, the total effect of the character is in counterpoint with the accepted notion of the type, just as it is in the case of Iago. The relation between the expected and the unexpected is more subtle with Othello than with Iago, because right to the end of the play we have a sense that Othello both represents and transcends the type, while Iago merely negates the type, merely runs counter to the idea which is firmly in the minds of the audience and of the other characters in the play. There is thus a delicate interplay between the satisfying and the frustrating of the audience's expectations—as, I believe, there is in Jacobean tragedy in the total picture of the world that is commonly presented by it.

Not only in the use of character-types and in the vision of the world presented but also in the turns of event and the theatrical technique there will often be a challenge of, a resistance to, the audience's expectations. We can think of Shakespeare ending his story of love and war in *Troilus and Cressida* by bringing Pandarus on to the stage, to announce the making of his will and to offer the audience his diseases as a legacy; of Wycherley, taking a cue from Molière and inserting into *The Plain Dealer* a mock-critique of *The Country Wife*; of Gay defying causation by producing a last-minute reprieve for Macheath; of Kleist suddenly (though perhaps only for a moment) transforming the romantic Prince of Homburg into a man begging for life on any terms; of Ibsen giving us, in many of his prose dramas, a contemporary setting and realistic speech along with a compressed action derived from ancient practice; of Shaw mocking his audience's expectations by adding an epilogue to *Saint Joan*; of Brecht bestowing on the drama the freedom, the amplitude, the lack of subordination of part to part, that are characteristic of the epic, while at the same time rejecting that Kind's high seriousness and involvement with

[1] See Eldred Jones, *Othello's Countrymen: The African in English Renaissance Drama*, London, 1965, especially p. 108.

character; of N. F. Simpson using the arbitrary connectives of the day-dream. The drama, when it is alive, will always be partially at odds with what the audience expects. At its most successful, it will end by convincing the audience that what it receives is what it has been wanting. But at the first performance of an aggressively new play this does not easily happen.

Nor can we be surprised that the first performance should arouse opposition, and a special problem is caused by the fact that plays rarely establish themselves after an initial failure.[1] We have seen how easy it is for the premature judgment to be formed and expressed while the performance is taking place. What we as spectators should aim at is a special doubleness of response in this regard. There is, we have noted, a double response to the performance as, on the one hand, an immediate presentation of actual life and, on the other, a rehearsed entertainment—as, in Samuel Johnson's words, 'a certain number of lines recited with just gesture and eloquent modulation'.[2] In a similar fashion we must allow our critical faculties to operate along two lines at once. We shall all the time be making judgments as we watch: they will be premature; they will be, or should be, subject to continual modification as the play proceeds; they may be communicated through laughter or through inter-act comment or through applause at a special moment of approval. But alongside this duck-shooting, as Swift has enabled us to call it, we must be holding aloof from giving sentence; we must preserve a measure of solitude as a defence against the contagion to which, as a member of a particular audience, we shall be subject; we must wait. This is not to say that there is to be no relation between the immediate, the growing, response during the performance and the response we arrive at as we remember the play. The one indeed will be the prime basis for the other, but an ultimate

---

[1] See below, 'The Dramatist's Experience', pp. 228–9.

[2] *Preface to Shakespeare with Proposals for Printing the Dramatick Works of William Shakespeare*, London, 1957, p. 27.

judgment is approachable only when not merely this play is held in the memory but along with it the whole body of drama that we know, the condition of the present time, the range of our experience as human beings. Then, but not till then, we shall be in a position to respond to the play-wright's full challenge. We shall still make mistakes, which we may come to regret. We have always to try to avoid the illusion of finality in judgment. But, in recognising from the beginning that the immediate and fluctuating response to the play during performance is not to be our last word, we can come a little nearer to the position of the true critic.

And in a measure that will cope with the problem that the play is never the same on two occasions and is to some extent affected by a particular audience's response. When we view it in retrospect, we shall be less influenced by the details, which are the things that fluctuate, and more concerned with the totality of the thing offered. Our judgment of the whole, which has thus come into operation only after a lapse of time, will be one which makes more sense to other people, who have seen the play on different occasions from ours. We shall even be able to exchange views of the play with someone who has seen a different production or has merely read it: our experiences will be in some measure different, but, unless one of the productions seen has been violently eccentric, we shall be discussing basically the same thing; the mere reader is, of course, at a major disadvantage, but at least the play exists for him more fully than a film exists for someone who has read only a published script. If we are resolute and searching enough, we can in retrospect bestow on a work of the theatre something of that very stability which, during the time of performance, it necessarily lacks. This means that the theatre can escape from its purely ephemeral condition: memory, consideration, comparison have made it truly public, a proper subject of discourse.

Chapter 11

THE DRAMATIST'S
EXPERIENCE

I T is appropriate to turn from the spectator to the dramatist, but of course there are other experiences than theirs
in the theatre—the actor's, the director's, the designer's, and
those of all the other people whose work goes to shape what
is done on the stage. However responsive the spectator may
be, however loyal to the author his fellow-workers in the
theatre, there can be no simple identity between what goes
on in a dramatist's mind and what anyone else experiences.
Yet I think most of us will agree that a good dramatic
performance is rendered more likely if everyone concerned—either as spectator or as theatre-worker—achieves
in substantial measure an appreciation of what the author
has been through. Even to-day, when we hear of 'playing
against the text' and of specious distortion of the text
being justified by an appeal to 'sub-text' (where any man's
guess can exercise a modish charm), most people would
admit to a certain interest in what the play has meant to
the author: we still put 'by' before his name on a theatre-
programme, thus ranking him along with novelist or poet.
Here, then, our concern will primarily be the kind of
experience a man goes through when he writes a play.

But first we should note how necessary the living
dramatic author is. There have been times when playhouses
have been heavily dependent on the writing of earlier
generations—notably in England in the eighteenth century,
in Greece after the Periclean age, in recent times in the
Kabuki and No theatres of Japan. These were not places
and times of theatrical sterility. Certainly, to take the
example I know best, the London stage of the eighteenth

H

century was a place of major acting where the inheritance from the previous century was in large measure consolidated. Although at that time many gross things were done in the playing of Shakespeare, it was then that his work was firmly established as the dominant part of a stable repertory. A theatre which disregards the drama of the past will be a theatre without standards, ever ready to dazzle itself with novelty, unaware moreover of when it is actually achieving novelty and when merely playing an old trick. But if the drama of the past is needed, so too is the new play. In the last few years we have been able to observe a parallel development in the conduct of the Royal Shakespeare Theatre at Stratford-upon-Avon and of the Stratford Festival in Ontario. In England it has been a bold and expensive undertaking to run the Aldwych Theatre in London along with the theatre at Stratford-upon-Avon, using some productions in both houses but including in the London programme a number of continental and American and new British plays. In Ontario the Avon Theatre with its proscenium-arch has been used for modern plays and for opera, which have thus been presented side-by-side with the Festival Theatre productions of Shakespeare's and other plays that seemed to lend themselves to the open-stage and the large auditorium. A dramatist as well as a spectator will gain from this linking of present and past: he has to see his own writing in the context of what the theatre at its best has already done. While he must at all costs avoid a mere imitating, a vain attempt to write as if he belonged to a society different from his own, he will know more fully what he is doing if his work has to take its chance along with a performance of *King Lear* last night or near-by, or the same company of actors are rehearsing or already playing in *The Way of the World*. If this thought is frightening, it is as well that a man should realise the size of the problem that faces him when he decides to write a play.

For the task is indeed a major one, even if it has its peculiar attractions. A writer of novels or poems does not often see someone actually reading what he has written: a dramatist not only knows that a group of people have worked together, learning his lines, planning stage-movements and settings which aim, or should aim, at being complementary to his words, but he can also go to the theatre and watch people listening to his play and responding to it in one way or another. Being thus engaged in a more obviously public activity than the novelist or the poet, he runs the greater risk of a social rebuff. An author may be laughed at if he appears on the stage at the end of a first-night performance, or he may merely sense an audience's indifference and have to meet the consolations of his friends, the disappointment of his actors. Still, it is not hard to find people who will run the risk, and far too many of them set out to write a play with insufficient thought for the number and the variousness of the demands that playwriting makes. There can be few things requiring more strenuous mental activity or a greater range of skills. Of these skills the first that we should consider is an ability to write dramatic language.

The text of a play is almost wholly made up of words that are to be spoken. The current interest in 'total theatre', with its use of dance and mime and song, has not substantially altered that. Indeed, so much does language dominate the drama that certain kinds of human behaviour cannot be directly presented on a stage. In actuality we may spend hours in silence and near-immobility, reading a book or sleeping or writing, or we may talk with a friend, sitting in our respective armchairs and arguing about the state of the world and the folly of the people we know. The stage may hint at such activity, showing a man at the beginning or end of his solitary reading or sleeping, or giving a quite brief sketch of a conversation with a friend. But the staple mode of drama is talk accompanied by purposive and illustrative action, and no soliloquy or even duologue must last longer than a quite small number of minutes. That means, of course, that all dramatic language is condensed, as we can see by comparing what is actually said even in the highly discursive second act of *Hamlet* with the actual range of talk that we commonly get into a day filled with a comparable series of encounters, or if we measure what Hamlet and Polonius and the rest say against one of the long conversations that Thomas Mann gives us in *The Magic Mountain*. In the theatre we have a mere impression of men talking together: it cannot, as the novel can, offer anything like a replica of the what and the how of real-life speech.

Yet the dramatist's language must not only be suggestive of ordinary human converse: it must also be good for an actor to speak, good for the audience to listen to. Here the actor is the best guide to what is needed. It is he who can tell us most authoritatively how this man's words are laborious to utter and that man's a delight. The stage is a public place and the spectators are not yet the author's friends: we shall forgive the people we know when they fumble for words and reach only an approximation to the thing they imagine or hope they are saying, but we shall not so easily forgive the man who thus addresses us from a platform. The stage may present an inarticulate man, or a man clumsy with words, but his very inarticulateness or clumsiness must have an authority, an economy, that are foreign to the actual. It should hardly be necessary to say this, but some reading of contemporary published and unpublished plays will show that the minor twentieth-century playwright too often thinks that his habitual approximations are good enough for the stage. Nearly a score of years ago Eliot thought the point worth making in his Theodore Spencer Memorial Lecture, *Poetry and Drama*, at Harvard:

> Whether we use prose or verse on the stage, they are both but means to an end. The difference, from one point of view, is not so great as we might think. In those prose plays which survive, which are read and produced on the stage by later generations, the prose in which the characters speak is as remote, for the best part, from the vocabulary, syntax and rhythm of our ordinary speech—with its fumbling for words, its constant recourse to approximation, its disorders and its

unfinished sentences—as verse is. Like verse, it has been written, and re-written. Our two greatest prose stylists in the drama— apart from Shakespeare and the other Elizabethans who mixed prose and verse in the same play—are, I believe, Congreve and Bernard Shaw. A speech by a character of Congreve or of Shaw has—however clearly the characters may be differentiated—that unmistakable personal rhythm which is the mark of a prose style, and of which only the most accomplished conversational- ists—who are for that matter usually monologuists—show any trace in their talk. We have all heard (too often!) of Molière's character who expressed surprise when told that he spoke prose. But it was M. Jourdain who was right, and not his mentor or his creator: he did not speak prose—he only talked. For I mean to draw a triple distinction: between prose, and verse, and our ordinary speech which is mostly below the level of either verse or prose. So if you look at it in this way, it will appear that prose, on the stage, is as artificial as verse: or alternatively, that verse can be as natural as prose.[1]

A man in real life, condemned to death and anxious for a reprieve, may get out the words 'Tell me now, I can face it.' Shakespeare in *Measure for Measure* gives us a man in that situation, but his condemned Claudio speaks in a different fashion when Isabella hesitates to tell him of Angelo's rigour:

> Why give you me this shame?
> Think you I can a resolution fetch
> From flow'ry tenderness? If I must die,
> I will encounter darkness as a bride
> And hug it in mine arms.   (III. i. 81–5)

Of course, Shakespeare tells us something of Claudio's character in this: the fantastic image hints at an inflation of his courage, an unsureness of his own strength; we are subtly prepared for his breakdown a few moments later, when he begs Isabella to let him live by paying Angelo's price. But, quite apart from characterisation, the lines

---

[1] *On Poetry and Poets*, London, 1957, pp. 72–3.

are instinct with the theatre. Men can speak like this only on the stage, and there they should speak like this. It may be done in prose, as when Valentine in *Love for Love* reproaches Angelica for her apparent over-lightness of heart:

> You're a Woman,—One to whom Heav'n gave Beauty, when it grafted Roses on a Briar. You are the reflection of Heav'n in a Pond, and he that leaps at you is sunk. You are all white, a sheet of lovely spotless Paper, when you first are Born; but you are to be scrawl'd and blotted by every Goose's Quill. I know you; for I lov'd a Woman, and lov'd her so long, that I found out a strange thing: I found out what a Woman was good for.
>
> *Tattle*. Aye, prithee, what's that?
>
> *Valentine*. Why to keep a Secret.
>
> *Tattle*. O Lord!
>
> *Valentine*. O exceeding good to keep a Secret: For tho' she should tell, yet she is not to be believ'd.   (IV. i. 634–46)[1]

Or as so often in Synge and O'Casey, where the words have a relish and the imagery a fecundity that even the Irish cannot be expected to master in their everyday talk. Throughout a play there must be a sense of fine speech, but there are moments when the audience will grow especially conscious of it: these moments when the words themselves 'Move us with conscious pleasure', as Wordsworth said of his early experience of poetry in Book V of *The Prelude*, are as much part of the theatre as the moments of striking action, the strong curtains of the 'well-made play', the outbreaks of violence and laughter and grief. That is to say, certain sequences of words will stand out as of a peculiarly highly-wrought character, and the audience's attention will be sharpened. Coleridge, without reference to the drama, proclaimed a similar phenomenon when he declared that a long poem cannot and should not be all poetry but should furnish an appropriate setting for the occasions when poetry

[1] *The Complete Plays of William Congreve*, ed. Herbert Davis, Chicago and London, 1967, p. 292.

was achieved.[1] Eliot, in the lecture I have already quoted from, echoed Coleridge in this, but seems wrongly to have deduced that the audience should not be aware of the point of transition:

> It is indeed necessary for any long poem, if it is to escape monotony, to be able to say homely things without bathos, as well as to take the highest flights without sounding exaggerated. And it is still more important in a play, especially if it is concerned with contemporary life. The reason for writing even the more pedestrian parts of a verse play in verse instead of prose is, however, not only to avoid calling the audience's attention to the fact that it is at other moments listening to poetry. It is also that the verse rhythm should have its effect upon the hearers, without their being conscious of it.[2]

He was right about the setting: all of a play's language must make it possible for the moment of high utterance to emerge without strain; but wrong, I think, in believing that an audience should not be aware of the special moment as in every way distinctive. It is an error that the contemporary theatre falls into with its productions of Shakespeare. Recently I have seen a *Troilus and Cressida* where the actor playing Ulysses spoke both the 'degree' speech and 'Time hath, my lord, a wallet at his back' as if he were anxious that we should not believe he was exerting any special power. The theatre, as the nineteenth century well knew, rejoices in special moments.

Yet it will not do for the playwright to cultivate mere flowers of speech, so as to make the audience aware that he has a 'literary' or 'poetic' strain. This was the bane of nineteenth-century verse-playwriting, and it had its analogues when Stephen Phillips or even Wilde (in his conscious attempts at 'serious' drama) laboured to keep English up. Shakespeare's language grows out of the common Elizabethan language, which profited from a widespread delight in the handling of words. Congreve's

---

[1] *Biographia Literaria*, Chapter XIV.
[2] *On Poetry and Poets*, pp. 74-5.

prose is the ultimate growth from a society that was anxious to show polish in its social encounters. Synge and O'Casey were privileged in belonging to a country where urban civilisation had not worn down the contours of words, making them into the indistinguishable small change of necessary traffic. The task in English to-day is more difficult. Even so, our theatre-language must be shaped and vivified and must never be afraid of the kind of larger utterance that for example Churchill gave us. In the theatre above all places, inhibition becomes tyranny. But if good dramatic language is to be written, it will not come straight out of the pen. There is much labour in writing anything: good plays do not write themselves any more than good poems.

Yet a dramatist is not concerned only with effective words. He must give us embodiments of imagined men and women in whom we can be interested. Shakespeare criticism to-day is inclined to minimise the element that character contributes to his plays—in some measure rightly so, for the last century was too ready to see him as principally offering us a portrait-gallery of notable persons. No dramatic personage is as complex as any actual human being, but some figures in some plays do give us the impression of real-life people, just as a dramatist may give us the impression that we are hearing words that might be spoken off the stage. What we can demand of the characters of a play is that they make us aware of human behaviour, even though we know that they are caricatures or simplifications of what people really are. And of course we have a natural interest in caricature and simplification: that is how we commonly think of those around us. But the picture of the people presented must be as vivid as, for example, a David Low cartoon was for the newspaper-readers who responded to it. And again a dramatist must have the actor in mind: it is the actor who must feel that the character belongs in his world and is worth presenting to his world. He will not mind an extreme degree of simplification if he feels that a point is being made through the character, but he should resent simplism, the way out for the man who courts the easy answer. The characters must be vivid for the duration of the play, as Face and Subtle are in *The Alchemist*. Sometimes, however, the dramatist will be as deeply conscious of the men and women he imagines as of those whom he knows most intimately in the world he lives in when he is not engaged in composition: then indeed it may be that his characters run away with the play. But that can also happen even when the figures of the drama are strictly limited to the world of the stage. Referring to the 'vividness of the descriptions or declamations in Donne or Dryden', Coleridge remarked that 'The wheels take fire from the mere rapidity of their motion.'[1] So it may be when dramatic characters, however simplified, acquire the special energy that may come through intimate involvement in a plot—provided always that the audience can feel some relation to the current of life. Supreme examples of this phenomenon are to be found in Goldoni's comedies.

A dramatist, moreover, is a man who can work with others—a man of reasonably mobile temperament, ready to see that what he has written may need to be modified when it comes to be realised in the theatre, that the actors may feel his characters as he has first presented them do not properly belong within an apprehensible world. His appreciation of his fellow-workers as human beings will be linked with his ability to imagine parts on the stage. Yet, working with others as he must, the dramatist needs to remain the author. It is his play that is being acted—no longer totally his[2] but something that, while preparations for performance are going on, he has a responsibility for: he must preserve a nice balance between possessiveness and abdication. All this implies an ability to belong intimately to a social group while remaining aware that he has had a moment of special vision to which he must be loyal. In this respect his relationship to his fellow-workers is emblematic of his relationship to the audience that the

[1] *Biographia Literaria*, Chapter XVIII.

[2] See above, ' "The Servants Will Do That For Us" ', p. 20.

particular theatre can expect to draw to itself. The dramatist, through the combined activities that take place on the stage, commits himself to an utterance which the audience as a group has to respond to, receiving it in substantial measure as their utterance too. The play as it comes to birth on the stage is a statement made on the audience's behalf, yet that most surely does not mean that the author is to see himself as a mere mouthpiece, as one giving to his audience what they already have in the forefront of their minds. He is not a pageant-writer who lets a group of people have their prejudices finely aired on an evening out. This, a matter already referred to in this book,[1] will have to be returned to when the dramatist's choice of theme is considered.

We have briefly noted the play's language and characters, but a word at least must be said about what is done by the characters as distinct from what they say. We have already seen that not all human behaviour lends itself to dramatisation. Moreover, what is playable has to be handled in a special way. The problem is partly again one of condensation and also one of continuous interest. The time-span of a dramatic performance may vary only within quite narrow limits, from the one-acter to what we think of as a whole evening's 'entertainment'. There are always those dramatists in the twentieth-century theatre who defy its social conventions, giving us *Back to Methuselah!* and *Strange Interlude* and *Mourning Becomes Electra* and *Le Soulier de Satin*; but Shaw never expected *Methuselah* to be played, O'Neill's long plays won stage-success in America only because of his special eminence there, and Jean-Louis Barrault had to abbreviate Claudel's play rather drastically before it could become theatrically viable. In other times different social conventions have operated: the tragic trilogy followed by a satyr play was required in Hellas, the cycle-play of late medieval years took up a whole day, as in the sixteenth century Lyndsay's *Satire of the Three Estates* could. But the

---

[1] See above, '"The Servants Will Do That For Us"', pp. 21–2; 'On Seeing a Play', pp. 207–11.

idea of a normal limit—from two or three hours in Shakespeare's time and ours to a day's length in other societies—
is still there. No one prescribes how many lines an epic shall
have (though the number of its separate books has been
a matter of critical doctrine) or how many pages a novel
shall run to. Dramatists have chafed at the limits imposed
on them: as early as the first publication of Webster's
The Duchess of Malfi in 1623, we read on the title-page that
the play-book contains 'diuerse *things . . . that the length of the
Play would* not beare in the Presentment', and in the Restoration and eighteenth century it was common for printed
plays to mark passages that were not spoken on the stage.

This sign of a tension between the dramatist and the
medium in which he works is something that we shall need
to explore further. But in so far as the play is to exist in
an actual theatre—and at least in an important sense that
is the place of its most manifest existence—there is not
much room for the development of surge and thunder, for
the detailed exploring of act and aspiration. Moreover,
within this span of time the attention of the audience must be
continuously held. The novelist can give us stretches
where we grit our teeth in a determination to get through—
either because what we have already read, in the book's
beginning or in other books by the same author, has
convinced us that patience will be rewarded, or because
those on whose judgment we rely have assured us that the
effort will be worth while. We often tell our friends that they
must not be put off by the first hundred pages of a novel,
as for example in *The Ambassadors* (if James is a new author
for the particular reader) or Malraux's *L'Espoir*. Indeed there
is a strong case for believing that the major novel depends
for part of its stature on its very *longueurs*: we must be made
to feel that time hangs while Hans Kastorp spends his years
on the Magic Mountain, while Russia undergoes its trial
during the drawn-out alternations of War and Peace. The
long novel that can be read too briskly is as quickly
evanescent, just as, in reverse, one remembers better the
language whose vocabulary and accidence and syntax have

taken the greater while to learn. But with a novel we can lose patience for an evening and yet feel impelled to take up the task again to-morrow. In any event, *The Magic Mountain* and *War and Peace* and *Nostromo* are not to be read at a sitting. In the intervals we recover breath and can make further demands on what stamina we have. But the play cannot be left: a moment's inattention is, for that particular performance, irrecoverable, and only a few addicts return to the theatre for a second attempt. In the novel and the epic there can be a boredom which is part of the magnificence of the major work; in the drama attention can be slackened occasionally at the dramatist's will—we can think of Edgar's account in *Lear* of how his father died—but only the dramatist of the highest rank should dare to risk this for more than a short while. It is a valid generalisation that throughout a play, or indeed throughout any spoken use of language, even in a lecture, there is fatality in wait for the writer who gives his hearers the occasion for inattention.

We sometimes hear the expression 'a captive audience', but there is no such thing. All of us know how often we cease to listen, and brood on everyday concerns or on matters that we feel are of more weight to us than the things done and said from the stage and the platform. It is surprising, perhaps, that anyone dares to risk offering a two-hour play or a one-hour lecture. Moreover, we write for such-and-such an audience at such-and-such a time. If we are successful, other audiences in other places and times may want to hark back: *Hamlet* has found its way, I suppose, to every country since it was first acted in Shakespeare's London. But the harking back will not normally take place unless the original impact has been strong. There are exceptions: *Troilus and Cressida* was probably acted only to a small group in the dramatist's life-time; *The White Devil* was an admitted failure when put on for the first time at the unsuitable Red Bull Theatre; *The Playboy of the Western World* could hardly be heard when the Abbey in Dublin tried to act it first; in our own time Harold Pinter's *The Birthday Party* caused puzzlement and some mockery at the Lyric, Hammersmith. These are splendid exceptions, and no dramatist should be so arrogant as to count on them as comforting precedents. If he does not make his impact now, when his first audience watches, it is unlikely that he will, for this particular play, have a second chance. So his action, the sequence of events that his words involve, must exert an immediate spell.

Words and characters and action are what the dramatist himself brings to the performance, and we have seen he must be in sympathy with the theatrical *milieu*, must be not merely content but needing to work with others, sympathetic too with the audience he is to address—though not as a suitor, not as one obsequious in their presence. He must also have some understanding of theatrecraft in general. His words and actions, his men and women, will be viewed in a setting that another man has designed or that is more or less permanent within the theatre his play is intended for. Sometimes indeed the idea of the stage-setting can be a formative element in the play's structure. Arthur Miller perhaps wrote *Death of a Salesman* in close contact with Jo Mielziner, the designer of the set, and the play took the shape it did because the setting suggested certain lines of development. It was a bare multiple-setting, corresponding well with the subdued language of the play and lending itself easily to rapid localisation and delocalisation. This does not mean that every dramatist needs to have a scene-designer at his elbow from the beginning, nor does it mean that every good set should be of the semi-constructivist kind used in this play. But it does suggest the strength that can come when at the outset the dramatist thinks of a particular setting as a live element in the growing work. O'Neill, always fascinated by any aspect of the 'unities' (even in his moments of flagrant truancy), is in love, one feels, with the scenes he imagines. The more fully they operate on the characters and their actions, the more he seems convinced of the validity of what is said and done. In *Desire under the Elms* and *A Moon for the Misbegotten* he shows first the exterior of his characters' home and then the interior: walls are

229

removed, so that we see the inside of the house while the outside is still in view, framing the action within. We approach by degrees to the hearth where they warm themselves, as Ibsen took us after an act's delay to the attic where the Ekdals lived. In *The Iceman Cometh* we go, in different scenes, from one end to the other of Harry Hope's bar, so that we have the experience, as habitual visitors, of wandering at will within the orbit of the play's action.[1] This is another matter than the laborious numbering of a room's doors, windows, pieces of furniture in an initial scene-description: O'Neill, one feels, does not set great store by such details, though he gives some sort of list of them to make the action clear; what more concerns him is the feeling, the general size and shape, of a particular place, which exercises its influence on what is said and done. I should also be prepared to guess that one of the first things that developed in Graham Greene's mind when he was writing *The Living Room* was the living room itself: it was not a mere framework within which the play could be conveniently acted; rather, it was the key-symbol of the play and the inevitable living-place for the frightened people we met there. Greene had first to imagine the house's last-but-one room to be touched by death, the room which, when death had in fact set its mark on it, gave to the surviving characters their last opportunity for an act of will.

The dramatist who writes for a stage which in itself provides the contour of the setting—whether Shakespeare's stage or an open-stage such as we commonly know to-day—still has a basic question to answer. Shall he, despite the manifestly anti-verisimilar platform he is to project his play from, attempt to convey a sharp sense of a particular place, or shall he set his play against a void, challenging his audience to see his action in the nakedness and universality of nowhere? Shakespeare did both, making us intimate with Elsinore and the island of Cyprus, keeping us away from

[1] For reasons of theatrical convenience, this was not done in the London production.

any firm sense of locality for the action in *Lear*. In *Lear* indeed he played the supreme trick on us, making us most aware of Dover Cliff, where no scene takes place. But when he chose, as Webster chose in certain scenes of *The Duchess of Malfi*, to make us feel truly at home in a particular locality, he could do it through dialogue and through the powerful association of certain characters with places they dominated. We know Elsinore and Malfi as closely as we know Hedda Gabler's room, cluttered as that is with the paraphernalia that a woman still near to us in time can accumulate and make her own. This kind of effect can be hinted at in a few words of scene-description, which, whatever mode of staging is being used, the play's director must take as his cue. Act I of *The Cherry Orchard* takes place, we are told, in 'A room, that has always been called the nursery':[1] that, as we read, is enough to put us there, is something for the actors to live up to. If a beginning playwright needs the advice, the imagining of a place of action is not the worst starting-point for one's composition. Even so, there are major plays that need the void, that would be damaged by over-precise localisation.

But an author will commonly feel he has more to do than show men and women speaking and acting in a particular setting. He wants to say something about our special concerns to-day—about Vietnam, or Czechoslovakia, or Biafra, or Chicago, or the London School of Economics, or about the death of God. It is indeed more than possible to start with a general or abstract idea, related to the nature of society or the cosmos or the political upheavals and follies of our time. Perhaps that was what motivated Shakespeare when he wrote *Love's Labour's Lost* and *Troilus and Cressida* and *King Lear* and *The Tempest*. But these examples show the indirectness of his method, show how his mockery and terror and puzzlement were rooted in his clear-eyed presentation of imagined men and women against an imagined background. A body of doctrine can manifestly

---

[1] So in Constance Garnett's translation.

inform a play, as doubtless with Aeschylus in the *Oresteia* and the *Prometheia*, with the medieval playwrights who told the story of the world as Christianity had bequeathed it to them, as to-day with those dramatists of China and the Soviet Union who have, on a generally lower level, promulgated their states' teaching. And even the most individual playwright will have a general point of view, an attitude to his immediate world that will condition what his characters do and say. But he should guard against mere preachment, particularly in the society in which we now live. We in the audience are a heterogeneous group, of differing attitudes towards the cosmos and towards society: as a group, therefore, we shall be quickly out of sympathy with an attempt to dramatise a sermon, ready to resent the play that blatantly tells us what we ought to think and that therefore sets up divisive tendencies among us. We shall be willing to accept doctrine only when we are caught up with the characters and action so deeply that they in themselves matter to us. My personal response was one of close involvement when I saw the Piscator dramatisation of *War and Peace* in London some years ago, because Natasha and Pierre and André were so fully alive on the stage and because the war's pattern was movingly realised; yet even on that occasion as I left the theatre another spectator remarked: 'That was a piece of tub-thumping, wasn't it?' Nothing dies so quickly as tub-thumping, whether on our side or on theirs. I think the annoyed spectator's dismissal of the Piscator *War and Peace* was wrong, though I may be, I think understandably, prejudiced. What I am sure of is that some verse-plays written some years ago and influenced by Eliot, for example Anne Ridler's *The Shadow Factory* and Ronald Duncan's *This Way to the Tomb*, depended on a mistaken notion of dramaturgy, reinforcing the converted perhaps, seeming intolerably naked to those unwilling to declare themselves immediately in favour of the ideas proclaimed. If our civilisation is worth anything, it is because it allows the possibility of difference—and, in any event, difference is what we have got. We have seen that the playwright is not a mere mouthpiece for his audience's already established ideas.[1] The direct enunciation of doctrine will at most win only an inert acceptance, but a spectator will not too strongly resent a point of view if he becomes interested in the people exhibited to him, if he feels compelled to watch what they do and listen to what they say. We may leave Graham Greene's *The Living Room* arguing on one side or the other, for the priest or the psychologist (though in the end we may believe that both surely were in some measure wrong), but we are fierce about it because the girl who is now dead has once been so thoroughly alive. It is well known that Arnold Bennett is said to have first glimpsed the idea of *The Old Wives' Tale* when he saw an old woman in a Paris café.[2] In its inception, the drama too may commonly be like that. On it the theme will grow.

Yet there is another matter we have to face. We have demanded much of the dramatist—an expertise in speakable and economical language, a vivid sense of human beings both as he imagines them and as he works with them, an ability to plan his action sharply and with continuous interest, a sympathy with his scene-designer or at least an awareness of what locality or its absence may mean to a play, a sense of theme without a habit of preaching. We have to recognise that he cannot in addition be Newton or Darwin or Freud or Marx or Wittgenstein. Arnold thought that poetry might take the place of religion for a world where Christianity had worn thin. So it may, but with the reservation that our poets and dramatists are not primarily thinkers. It is true that the Greeks and Shakespeare can enunciate for us the profound commonplaces which may be guiding-lines for how we live, as in nineteenth-century England Wordsworth did, as Goethe has for Germans since his time, as Dante has for Europe. But guiding-lines are not total systems, and even total systems get frayed or modified. Among Shakespeare's contemporaries in the drama we

---

[1] See above, '"The Servants Will Do That For Us"', p. 22.
[2] Everyman's Library edition, reprinted 1939, p. vii.

can find perhaps only in Chapman the sort of effectively generalised utterances that he himself gave us in certain places in *Lear* and *The Tempest* and *Troilus and Cressida*. Marlowe and Jonson and Webster give us rather a sense of the wisdom of the writer in that he saw that a particular person in particular circumstances ought to say this or that particular thing. And that much we can demand from a dramatist. Reading recently in an anthology some well-known plays of the last hundred years, I have come upon two passages that may be quoted here. In Arthur Schnitzler's *The Lonely Way* there is a man who has denied a child to a woman he loved, talking with a man who has had a son by a woman he abandoned:

Julian. ... there's no use arguing with you because you see you've never loved anybody in your life.

Sala. Perhaps not. And you? Just about as much as I have, Julian ... To love means to live for another. Mind you, I don't say that's a desirable condition, but I'm pretty sure that neither of us has ever experienced it. What do people like you and me really know of love? We've had our fling—we've known tenderness and deceit—yes even passion—but none of that's the real thing ... Have we ever made any sacrifice that didn't gratify our senses and our vanity? ... Have we ever hesitated to deceive decent people, or lie to them, for the sake of an hour of happiness or even lust? ... Have we ever really risked anything—I don't mean on the spur of the moment ... but actually to benefit a woman who had given herself to us? ... Have we ever given up anything except when it was the wisest thing to do? ... Is there anyone from whom we have a *right* to demand gratitude? I don't mean a string of pearls or money or cheap advice, but a part of our being—an hour of our existence, that we really gave them without any thought of getting something in return? My dear Julian, we left our doors wide open and everything we had was there for the world to see—but have we ever really been free with what was ours? You as little as I. We can shake hands on that all right. I just don't grumble as much as you—that's the only difference between us.—But you knew all this, didn't you? Just as well as I did. And how could either of us help knowing it? All the talk in the world couldn't change

234

it. Maybe others don't see our stupidity and rottenness—but we do. In our innermost souls we've always known, haven't we? But it's getting cold, isn't it? Let's go in.[1]

And in Arthur Wing Pinero's *Mid-Channel*, a rather melo-dramatic 'problem play' ending with an obvious echo of *Hedda Gabler*, there is a girl speaking, thinking she is going to celebrate her engagement to the man she loves, though we know it will not lead to any sort of happy marriage:

> *Ethel [kneeling upon the settee on the left, her elbows on the back of it, gazing into space].*   Mother—
> *Mrs. Pierpoint.*   Eh? [*Receiving her fan from Leonard*] Thank you.
> *Ethel [slowly].*   Mother—this is going to be an awfully happy night.
> *Mrs. Pierpoint.*   I'm sure I hope so, my darling. It won't be my fault if it isn't—[*tapping Leonard's shoulder with her fan*]—nor Leonard's.
> *Ethel.*   Ah, no; I mean *the* night of one's life, perhaps.
> *Mrs. Pierpoint.*   Oh, I trust we shall have many, many—
> *Leonard.*   Rather!
> *Ethel [raising herself and gripping the back of the settee].*   No, no; you don't understand, you gabies. In everybody's life there's one especial moment—
> *Mrs. Pierpoint.*   Moment?
> *Ethel.*   Hour—day—night; when all the world seems *yours*—as if it had been made for *you*, and when you can't help pitying other people—they seem so ordinary and insignificant. Well, I believe this is to be *my* evening.[2]

These are the words of sharply imagined people, Sala and Ethel, not unusually percipient but emotionally involved in the moment that is embodied in their speech and achieving a higher degree of articulateness than is normally theirs. The audience will remember such words as coming from people of their own kind and as simultaneously breaking

---

[1] *Representative Modern Dramas,* ed. C. H. Whitman, New York, 1936, p. 187. The translation is by Julian Leigh.

[2] *Ibid.,* p. 705.

through the customary bounds of reticence. Schnitzler and Pinero are not fashionable dramatists to-day, but here the feelings and ideas suddenly sharpen our sense of the people speaking. A dramatist will need to make us feel the capacity of people in this way if in the theatre he wants us also to care about his views on society and the cosmos.

If, as I think Piscator achieved in his version of *War and Peace*, the people in their acts and words and setting come across to us as people, so that the dramatist makes us at one with what he has imagined, the immediate end will have been reached. We respond to the people acting and suffering; we shall also, whether we like it or not, see the author's general view of things as viable, demanding respect, and shall be the more ready to scrutinise our own customary attitudes. And for him, too, there are great dividends to be had from the theatre: he is apart from his audience and his fellow-workers, yet he is one with them; he can appreciate the special skills of his actors (keeping aloof from jealousy), welcoming the audience's approval of a particular exercise of eloquence or movement or stance (as is notably the custom to-day in Kabuki performances); he can see how the very stylisation that theatre demands can release emotion from the flux in which it is customarily adrift (a release, indeed, that all art bestows); he can recognise the special mimesis that the living actor confers on his text. Civilisations have existed without drama, but it is a special privilege to live in a society where the actor gives to the playwright and the audience his peculiar incarnation of the words of a script.

Yet not every author can demand or has demanded this. There are those who have written in indifference to the current theatrical requirements of their society; and these dramatists have included Milton (for *Samson* though not for *Comus*) and Byron and in at least one instance Claudel, whose *Le Soulier de Satin* had to wait for a play-adapter, and perhaps even Büchner, whose *Wozzeck* firmly established itself on the stage only when Alban Berg made it into an opera. If we ask why the dramatic or semi-dramatic form has none the less been chosen, the answer probably lies partly in the prestige of ancient tragedy (which, of course, was often written without substantial hope of production) and partly in the impersonality of the dramatic mode: in using a form where, apart from stage-directions, every word is assigned to someone other than the author, one can feel one has the advantage of remoteness both from the reader and from the imagined characters. The idea of myth (if that too is wanted) seems thus more attainable. No valid objection can be made to use of the dramatic mode without relation to the theatre. It was a way of writing which the nineteenth and early twentieth centuries especially cultivated, and the English products in this mode included *Prometheus Unbound* and *The Dynasts*. It is a different and damaging matter when the writer hankers after a theatre without considering what performance in his place and time properly requires.

On the other hand, the poet-playwright—and every playwright, because he must use language with full deliberateness, has a close kinship with the poet—is of necessity a man whose total range of expression cannot be generally apprehensible in the theatre. From spectator to spectator there will be variation, dependent on the fluctuation of attention we have already noted, on habit of mind, on degree of sensitivity to language, on familiarity with the author's special idiom, on familiarity indeed with the particular play. And no spectator, however responsive, however persistent, can achieve a total intimacy with a major dramatic text. We all have the experience of hastily jotting down notes when we are seeing plays we thought we thoroughly knew. In the last thirty or more years there have been critics who assure us that the most important thing in highly-wrought drama is poetic imagery, yet no one could pretend to master a play's imagery when first seeing it in the theatre. That the critical preoccupation with imagery has on occasion become too exclusive would, I expect, be now commonly accepted. To swing to the other extreme and to say that such imagery, because not generally or immediately

available, is on the play's periphery would be as mistaken
as to find in it the one object of critical concern. Yet I think
we should in general resist the temptation to set up a
dichotomy between the play as acted and the play as read,
as if the dramatist while in the process of composition were
trying to address himself to a viewing and a reading public
simultaneously—even though the evidence I have
mentioned from *The Duchess of Malfi* might lead us to that,
even though *Hamlet* was probably never played in its entirety
until the twentieth century. Rather than separating the
reading public from the theatre public, we should think
of two distinct kinds of theatre which the dramatist will
have in mind. During the whole process of composition he
will normally look forward to his work being produced
in a real theatre, with actors (perhaps particular actors)
generally competent, with a director and his technicians
doing what they can to realise the proposed action and scene
and atmosphere, and with an audience of such intelligence
as the place and time of performance may give one reason
to hope for. But simultaneously he will be seeing the play
as acted in a theatre of his imagination, with actors ideally
fitted to their parts, with a director wholly understanding.
This is quite a different thing from envisaging a solitary
reader working his way through dialogue and stage-
directions. In this imaginary theatre the spectator will
be ideal too. His memory of the play's totality will be equal
to the author's own: he will pick up every reference, will
notice every parallel or contrast, will respond to the intellectual content and emotional envelope of each image or
reference as it is used. A dramatist who has only this ideal
theatre in mind will be unactable and often almost unread-
able, as many English verse-dramatists of the nineteenth-
century were. A dramatist who never thought of this ideal
theatre would give us only what easily lies within an actual
audience's compass, within their unstretched minds. Only
by the impingement on one another of these two theatres
in the playwright's mind can we have drama that is both
actable and supremely worth acting. It is the critic's

ultimate responsibility to try to put himself in the position of the ideal spectator in the ideal theatre which a particular dramatist has had in his mind. But the critic must also remember the importance which the actual theatre has had for the dramatist: he must never think that he is adequately criticising a play when he is thinking of it only in reading terms.[1]

When this double conception of theatre functions effectively in the author's mind, drama will come into existence that lives as acted and at the same time makes us aware of a dimension in its being that no performance can properly realise. That is the type-condition for major drama, but we have also to recognise that, quite apart from plays that are merely incompetent or are designedly unsuitable for the actual playhouses of their authors' society, there are also plays written by dramatic masters which nevertheless almost or perhaps totally defy tolerable production. Strindberg, who himself wrote some such, said in his *Open Letters to the Intimate Theatre* that for some dramatic writings we must fall back on reading as a more trustworthy guide than the inadequate performance which would be all that any theatre could offer:

> An actor cannot do without an author, but, if necessary, an author can do without the actor. I have never seen Goethe's *Faust* (Part II), Schiller's *Don Carlos*, or Shakespeare's *The Tempest* performed, but I have seen them all the same when I have read them, and there are good plays that should not be performed, that cannot bear to be seen.[2]

Here Strindberg comes rashly near to echoing Aristotle, who thought that a play might be as well read as seen.[3] But

---

[1] This matter, in particular relation to dramatic imagery, was referred to in my article 'Dramatic Imagery: Some Comments on its Range and Availability', *Critical Essays on English Literature Presented to Professor M. S. Duraiswami*, ed. V. S. Seturaman, Bombay, 1965, pp. 214–20.

[2] *Open Letters to the Intimate Theatre*, tr. and ed. Walter Johnson, Seattle and London, 1967, p. 21.

[3] *The Poetics*, Chapter VI.

my conjecture is that he is right about *The Tempest* (which I have seen more than once), and there may be other Shakespeare plays that approach a condition of resistance to production. That back-stage it is thought unlucky to quote from *Macbeth* is, I believe, a piece of evidence that this major tragedy has for long caused many actors' hopes to founder: certainly it seems to come nearer to defying performance than any other of Shakespeare's tragedies. But one has to be a very good dramatist indeed to defy the actual theatre in so triumphant a fashion as this. For the merely very good dramatists, and for the greatest dramatists in the bulk of their work, the twin conceptions of actual and ideal performance will live fruitfully together—always in tension, always reciprocally stimulating.

Theatre is ritual, and ritual implies belief, and what do we believe in? The implied situation could, it appears, mean death to theatre, yet theatre goes on—and not merely as a cultural gesture to the past, not merely as an entertainment for a large group which is prepared on occasion to lull itself with the idea that an outworn tradition is yet living. This is no time for a valid mass-theatre, because the masses, having (as masses) no tolerable belief, have no appropriate ritual. They can merely, and remotely, watch what television coldly offers. But this is most surely a time when a minority among us needs the theatre. All the major established influences to-day are in the direction of conformism: most people listen to the same radio, watch the same television, read almost identical newspapers; in the United States and the United Kingdom and some at least of the countries of Europe they vote for candidates who, in the major concerns, come near to echoing each other; we swallow, for the most part, the same bromides. This is not to disregard the existence of 'minority' radio and television programmes, 'minority' newspapers; but they are generally unknown to the majority, and can offer no ritual. Yet, with all the current drives to conformity, there have never been greater divisions between us, between young and old, between black and white, between West and East, between haves and have-nots. In some places government propaganda tells us we are at one within our frontiers, yet we fear the man at the garden-gate, as Harold Pinter has reminded us in his play *A Slight Ache*. Simultaneously we are driven together by social forces operating from established positions, and driven apart, it appears, by ever-growing prejudice and fear. Nevertheless, some of us are truly at one, perhaps in our very scepticism and discouragement, believing only (and with some difficulty) that life is worth living, worth thinking and writing about, worth examining when presented before us—believing, too, that people as people are worth knowing. We do not go to the theatre or to any art for a panacea, but in a theatrical context we have a sense of oneness with those of like minds, like minimal beliefs. There we respond to men and women acting and suffering as men and women must: the oneness is non-aggressive yet deeply attached: we sense that the dramatist and the players and all those in the theatre with us are anxious for what union can be achieved in so divided a world. Only in the theatre can we get so immediate a sense that people are, in the present as in the past, imperfect, aspiring, despairing, grabbing, fumbling, and difficult to know. Like ourselves, in fact. Only there, too, have we a sense of a minimal ritual-act still possible for us. It is most surely worth writing a play, or seeing one, if thus we can strengthen our knowledge of each other, our possibility (within a minority group) of coming together.

INDEX

242

INDEX

INDEX

245